International Collaboration in Civil Aerospace

D1744153

To my family

International Collaboration in Civil Aerospace

Keith Hayward

Research for this book was supported by a grant from the
Nuffield Foundation, England.

 Frances Pinter (Publishers), London

First published in Great Britain in 1986 by
Frances Pinter (Publishers) Limited
25 Floral Street, London WC2E 9DS

British Cataloguing in Publication Data

Hayward, Keith
  International collaboration in civil aerospace
  1. Aircraft industry
  I. Title
  338.4′762913334   HD9711.A2

  ISBN 0-86187-607-5

Typeset by Computerset (MCL) Limited, Ely, Cambs.
Printed by SRP Ltd, Exeter.

Contents

Preface and Acknowledgements   vi

Abbreviations   ix

Introduction   1

1. Technology, the Market and the Producers   17

2. Airbus Industrie and the Airbus Programme   52

3. Boeing, MDD and Collaboration   93

4. Cooperation and Competition in Aero-Engines   125

5. Subsidies, Sales and the State   157

Concluding Observations   193

Index   213

Preface and Acknowledgements

In the course of preparing an earlier book on the British civil aircraft industry, a senior industrialist told me that building commercial aeroplanes was like owning newspapers, 'a lot of fun, but you can lose an awful lot of money very quickly'. Certainly the history of European civil aerospace has undoubtedly seen a considerable amount of (largely public) money spent in pursuit of an elusive market breakthrough. That being said, losing money on civil aerospace is not just a European disease. Americans have burnt their fingers as well. Lockheed and Convair in the 1950s, can attest to the observation that civil aviation is a 'heartbreak market', (though Convair did compound its risks by getting involved with Howard Hughes). However, the Americans have argued at length that their risks have been private risks, and that the European taxpayer has, perhaps, paid rather too much for the privilege of having a civil aircraft industry.

As this book seeks to show, the issues raised by civil aerospace development are far from being a simple matter of state-supported companies versus private enterprise. The structure and the dynamics of the civil aerospace industry comprise a complex of technological, commercial, industrial and political factors which defy simplistic nostrums and categorisation. There is also a wider context to the perhaps narrow concerns of a sectoral analysis. Relations between the United States and other members of the 'Western Alliance' are not as cordial as they once were. Although there is still a large degree of consensus about fundamental security issues, misunderstandings, competition and mutual suspicion are eating away at the fabric of interdependence, and civil aerospace is emerging as one area of dispute and controversy. This book, therefore, also intends to examine the emerging points of dispute and conflict in a major industrial sector where American commercial domination is being challenged by a strong European response. Inevitably, in a field where developments are so fluid, there are many loose ends; certainly, the rapidity of technological and commercial change is such as to make this no more than a 'snapshot' of the problem at a particular point in time. Nevertheless, to wait for a clearer future, when judgements can be made at leisure is a typical academic failing. If this book has to be, therefore, a partial account, it will indicate the lines of debate, and will provide a platform for further work.

Finally, I must thank the many people who have helped me to complete the research for this book and to educate a non-engineer about the complexities of civil aerospace. Many have asked to remain anonymous, and I must respect that request. However, I would like to express my deepest gratitude for the welcome and assistance I received from industrialists and officials in France, Great Britain and West Germany. I must apologise to American readers for the distinctly European orientation of the interview material, but I would like to thank those US companies who replied to my written requests for information. I would like to express my gratitude to the trustees and officials of the Nuffield Foundation for providing me with a grant which enabled me to interview European officials and industrialists. I can mention a number of individuals. First to Roger Williams who yet again battled through a complete draft to engender some clarity of thought in the final product; to Trevor Taylor and Roger Tooze for their invaluable observations and comments; finally, my father-in-law, Laurence Corbett for his skills as proof- reader and stylistic conscience. The final responsibility for any misjudgements, errors and omissions is mine.

List of Abbreviations

AWST	*Aviation Week and Space Technology*
BA	British Airways
BAC	British Aircraft Corporation
BAe	British Aerospace
B-Cal	British Caledonian
BSE	Bristol Siddeley Engines
DTI	Department of Trade and Industry
EC	European Community
ECGD	Export Credit Guarantee Department
Exim	Export-import Bank
GATT	General Agreement on Tariffs and Trade
GE	General Electric
GIE	Goupément d'Intérêts Economique
HoC	House of Commons
HoL	House of Lords
HSA	Hawker Siddeley Aviation
IAE	International Aero-Engines
IATA	International Air Transport Association
JAEC	Japanese Aero-Engine Company
JCAC	Japanese Commercial Aircraft Company
MDD	McDonnell Douglas
MITI	Ministry of International Trade and Industry
MoU	Memorandum of Understanding
P&W	Pratt and Whitney
Rolls	Rolls Royce
SBAC	Society of British Aerospace Companies
UDF	Unducted fan (propfan) engine

Introduction

The international civil aerospace industry

In the words of Thomas Bacher, Boeing's International Business Director, civil aerospace is 'one of today's most visible and highly publicised industrial topics'. To him, its technical challenges, the high risks of development and sale and the traditional glamour of flight 'combine to make it one of the most fascinating business endeavours of our time'. [1] Civil aerospace is certainly a hard and often ruthless arena. As one observer notes:

'The business of making and selling commercial airliners is not for the diffident or faint of heart. It is remarkably difficult and, by anyone's standards, intensely competitive. There are few industries that consume as much or more capital; certain others rely as heavily on quantities of highly skilled personnel; probably no other is involved with as many advanced technologies.... But what really sets the commercial airplane business apart is the enormity of the risks as the well as the costs that must be accepted; they create an array of obstacles to profitability, hence viability, which discourages all but the bold and the committed...the process itself is exciting and the rewards if attainable, are high and include power and influence on a world scale. Hence many have tried, few successfully.' [2]

It is also quite distinctly, an international industry. The market for its products is global, and, increasingly, the development and production of civil aircraft and aero-engines is based on international collaboration. It is not, however, characterised by the operation of multinational companies. The internationalisation of production and distribution in civil aerospace is largely based on nationally-based firms linked either through licensing, sub-contracting, or through full risk-sharing partnerships.

This apparent paradox is explained almost entirely by the industry's close identification with the state, and the perception of its strategic importance to national technological and industrial interests. In the first instance, this is a result of the aerospace industry's overall contribution to national security, and the historical connection between civil and military development. This was partly a natural consequence of the rise of air power as a tactical and strategic instrument of war. But since 1945, in the United States, Great Britain and France, aerospace has also become a major employer, especially of highly qualified personnel, and is well-regarded as a high value-added, export-orientated industry. Civil aerospace usually makes up a relatively small proportion of the various national aircraft industries, although in some cases, and for a few companies, it can rise to as much as two-thirds of output. However, civil aerospace can have a significant effect on a nation's economic performance. For example, one survey ranked commercial aerospace third in the list of US exports, with a surplus in 1985 of $2.2 billion. Every $1 billion in export sales generated 48 000 man-years of employment and added $2.2 billion to the US GNP. [3]

State involvement in civil aerospace can take many forms. Such are the costs and risks associated with civil aerospace, that outside the United States, the state has been directly involved in the financing of civil programmes. Even American firms have received indirect help for civil development from their government, primarily through defence contracting. Governments can apply pressure on domestic customers to buy the products of the national aircraft industry, or an airliner which contains indigenously produced components. All governments have actively supported the export of civil aircraft and engines through the provision of export credit and other inducements. The place and legitimacy of state involvement may well be questioned, but its existence and impact cannot be denied. As Thomas Bacher notes, 'while governmental actions do not alter basic economic laws (of civil aircraft production), the future of the industry may well be influenced by governmental roles and policies...'. [4]

The political significance of civil aerospace has increased in recent years as a direct result of the emergence of real competition between the United States and Western Europe. Since 1945, the industry has been dominated by the United States; of all jet airliners delivered since 1958, over 87% have been built by

American firms. American manufacturers still command the bulk of world civil aerospace sales; in 1985, American companies had over 62% of the outstanding orders for new aircraft. Up to the mid-1970s, European manufacturers had made little impact on the market. Only continued support from governments, often for social and broader economic reasons, kept civil production alive. However, with the growing success of the European Airbus, the Americans have become increasingly concerned that their domination of civil aerospace is facing a sustained and long-term threat. This is seen as a direct consequence of European subsidy policies, and represents a political as much as a commercial challenge; 'in the presumed determination of European governments to sustain Airbus Industrie and to promote its fortunes in the world markets, worried Americans see a unique national asset—the commercial airplane industry—being whittled down, just as other of America's prize industrial assets have been depleted by foreign competitors'. [5]

Technology and the international political economy

Developments in the civil aerospace industry, and the political controversy that has followed in their wake, form part of a wider context of industrial and technological concern currently affecting relations between Western industrial nations. Technology based industries such as civil aerospace, are rightly assigned a significant place in the international political economy. [6] Technology, and the rapid pace of technological innovation, is one of the most dynamic forces for change in international relations generally, and through its impact on the industrial base of the modern state, it has a pervasive effect on a nation's long-term economic performance. As Roger Williams puts it, 'the development and exploitation of technology tends to change power relations, whether these are social, economic, political or military'. [7] The effect of technology on the international political economy reflects the interaction of two pressures, 'supply push' and 'demand pull'. The former tends to shape political and economic options through the continual availability of technological products, largely the result of institutionalised research and development conducted by either public or private actors. The latter creates requirements for technology, or helps to shape 'the stream' of research and development through the judgement of private and public decision-makers as to what the 'consumers' of technology, society at large, corporate bodies, the government and the military, will be prepared to buy. The product of this interactive process has had 'ramifications for the international political economy (which) have themselves been profound'. Technological change has, *inter alia*, helped to account for the growth and power of the multinational corporation, the differential performance of broadly similar national economic and industrial systems, collaboration between states and other actors aimed at spreading the costs and risks of some technologies, and the triggering of a broader political and social sensitivity to the hazards and dangers of technological innovation. [8]

Technology clearly has had a decisive effect on the evolution of the international system. Historically, the interaction between technological competence and political power has been most evident in military affairs. Even in pre-classical antiquity, 'a series of important changes in weapons-systems resulting from sporadic technical discoveries...sufficed to change pre-existing conditions of warfare and army organisation'. These, in turn, induced 'far-reaching social and political upheavals'. [9] The industrial revolution led to the first systematic exploitation of technological innovation to enhance military power by the major European powers. Indeed, the origins of the modern 'military-industrial complex' can be found in the close cooperation between the state and technically specialised companies which appeared in France, Britain and Germany towards the end of the nineteenth century. [10] Two world wars, with increasingly complex technical demands, confirmed the importance of technology as an integral part of military prowess. Since 1945, with the onset of a technologically-based arms race between the Superpowers, the significance of technology in structuring power relationships has become self-evident.

Industrialisation also broadened the meaning of national power. The capacity to sustain a continuous cycle of technological innovation became a vital element in national economic growth. As a result, extensive government support for technology has become a standard feature of public policy throughout the developed world, and of increasing concern to underdeveloped states. Moreover, a specific group of industries and technologies, which since the Second World War has usually comprised aerospace, electronics and nuclear power, has been defined as 'leading', or 'strategically significant', possessing a wide-ranging ability to promote improvements in the overall technological competence of national industrial systems, and to act as a catalyst for economic growth. The definition of a strategic industry has varied with the appearance of new technologies. Shipbuilding and other heavy industrial technologies once had this status. Nuclear power, especially fission-based technology, seems to have lost some of its appeal. By the same token, a new set of 'vanguard technologies' is emerging; for example microchip production, information technology and the bio-technologies. But the same principles apply, 'it has been proposed that, in high income countries at least, technological change in these industries drives economic growth more generally and that government policies should explicitly aim to facilitate the progress and competitiveness of these industries'. [11]

These 'leading industries' share a number of basic features. They operate in environments of high technological and market uncertainty. Investment in specific, large-scale and extremely expensive projects is a gamble, compounded by the long lead-times inherent in the research, development and production of a complex technological product. Second, technological change in these sectors is not mono-linear, but affected by a series of interdependent improvements in a range of technologies. This suggests two broad approaches to the management of technological innovation. The first favours a gradual evolution of relevant technology, perhaps across a broad base of related areas. This tends to favour the established manufacturer who has perhaps already written-off the high cost of developing an existing technology. But the nature of technological change might also bring a sudden, revolutionary and unpredictable lurch in the state-of-the-art, which could support a second, but higher risk strategy, aimed at exploiting the new technology in a single bold move. This could render the competition's products obsolete overnight; but it is fraught with greater technological and financial uncertainty. It may be an advantage to be first into the market, but equally, it may be better to 'second strike' the opposition, and to benefit from any lessons which may have been painfully learnt by a pioneer. This, as we will see, represents one of the classic dilemmas of civil aerospace development. The broader phenomenon of the 'latecomer' economy, such as Japan, learning from, and then surpassing, established industrial states, is well known. Significantly, Japan has now begun to adopt a 'leading edge' strategy in the face of competion from newly industrialising states, and again, one of the chosen areas is civil aerospace.

Finally, and perhaps inevitably, for governments determined to encourage technological innovation, there is no magic formula to guide either the strategy or the tactics of public policy. The tendency of some governments to throw good money after bad, locked into long lead-time technology, with its retinue of employees, and external and internal lobbists, is well known. On the other hand, and individual fiascos notwithstanding, the symbiotic relationship between the state and high technology is undoubtedly a force of fundamental importance in building the power of nations. But for the smaller state, with limited resources, the dilemma of a 'radical' versus a 'conservative' strategy is particularly acute. It cannot, perhaps, afford the broad-based approach, but neither can it afford to choose the wrong, or ill-judged 'vanguard' technology.

The comparative economic advantage of a developed state, therefore, has been seen increasingly to rely upon investment in a few, capital intensive, technologically-innovative industries. However, unequivocal evidence to support the vanguard technology thesis has proved hard to find. Indeed, several critics of government expenditure in these technologies have argued that their share of national research and development resources is largely an accident of the Second World War. Its continuation is in part explained by the political and institutional power of the vested interests which promoted nuclear power and aerospace in the United States and some Western European countries since 1945. According to this school of thought, an equivalent, if not a greater return might have accrued from more broadly-based investment in general economic and technological capability. [12]

Despite such scepticism, the 'leading industry' thesis, explicitly or implicitly, has achieved a strong hold over political perceptions, and has helped to shape many national industry and technology policies. The collective experience of the five major OECD countries, the USA, Great Britain, West Germany, France and Japan, shows that governments have equated their nations' relative standing in the international hierarchy with success in these sectors. Zysman's study of governmental intervention to encourage growth and industrial adjustment in these states demonstrates that there has been a pervasive concern to protect and to enhance a national technology base. Moreover, the most favoured technologies have tended to be electronics, aerospace and nuclear power. Some governments and industrialists may adhere to the myth that a free market acts as the best and normal mechanism for resource allocation. In reality, all have made a creative effort to 'plot the course of national industrial development'. There are national differences in style, points of departure, and the relative success of government intervention in each state, but the ultimate goal remains much the same, the development of national power and welfare through the promotion of key, technology-based industries. [13]

Technological issues, and the pressure of technological imperatives have also played their part in the evolution of the Western alliance system. This system was born as a result of security politics, and is still underpinned, even if with diminishing intensity, by shared perceptions of ideological and military threats. But from the outset, the Western powers have also been tied by a nexus of economic, industrial and commercial linkages. Although this was never a smooth nor equally beneficial process, the general trend seemed to be one of increasing interdependence across a spectrum of issue-areas. The technological dimension was regarded as being particularly important in forcing the pace of interdependence. Henry Nau, in his analysis of cooperation in nuclear power, reviewed some of the more ambitious prognoses for the future development of this system. Some suggested that there would be a steady growth of transnational networks and cooperative programmes amongst western states, particularly in Europe, which would, in turn, require a 'modification of past concepts of national independence and sovereignty'. [14]

For his part, Nau was not convinced that this was conceivable, let alone likely. Although he conceded that technological imperatives, 'supply push', to use Williams' terminology, to a degree shaped political choice, he also noted the importance of political factors in determining the context in which technological decisions were taken. More important, he analysed the strategic consequences of technological innovation, particularly where the United States, through deploying its greater resources and by achieving broad-band progress in a number of technologies, was able consequently to derive extensive economic and political influence from its relative technological superiority. The effect was to reinforce, rather than to dilute national power disequilibria. In short, although there was bound to be an interaction between 'political' and 'technological' variables, the former tended to outweigh the latter in the long term. Rather than technology helping to create new patterns of global behaviour through transnational linkages, in fact 'transnationalism may conceal an old pattern of national independence and international imperialism'. [15] The result was to create resentment and suspicion on the part of the technologically less advanced. In particular, it led to a growing European determination to resist and to match American technological 'imperialism'. [16]

Nau's 1974 study identified the reactive nature of European policies, a reaction that is to the perception of US technological dominance within the Western alliance system. This view of an inherent American advantage was supported by J. G. Ruggie, who suggested that 'particularly in the early post war years, but well into the 1950s and even the 1960s, American technological hegemony defined an order of relations within which others had to find their place'. [17] The US, as a 'leader state' in terms of technology, had derived considerable political and economic advantage over the rest of the western system. The literature and rhetoric of the 1960s, with its emotive language of an 'American Challenge' and 'an industrial helotry', found a receptive audience, and stimulated many political initiatives aimed at overcoming both national and regional shortcomings. At the time, various explanations for US technological dominance were advanced, including some which, quite rightly, identified fundamental socio-cultural faults in the ability of some European countries to adapt to the demands of a rapidly changing international technological community. In the final analysis, however, it was hard to escape the conclusion that, these weaknesses notwithstanding, the sheer size, integration and dynamism of American public and private markets for advanced technology provided the essential reason for Europe's comparative backwardness.

Part of the diagnosis of Europe's problem was, of course, a collective failure to make the best use of the region's evident technological and scientific resources. There was a distinct and repeated history of individual invention vitiated in the long term by inadequate national markets, limited capitalisation and intra-regional competition. A logical prognosis was that European-wide technological collaboration would help to overcome these national problems, and would place Europe on a more equal and competitive footing with the US. Despite some success with joint ventures, not only was the surge in collaboration during the 1960s flawed by the inherent difficulty of establishing transnational programmes, but national strategies often ran counter to collaboration. There were difficulties associated with national commitments to cooperative programmes. Some promising ideas never got started, others collapsed early in the development phase. Industrial inefficiency caused by the application of '*le juste retour*' to work and technology sharing was a persistent problem. Even more telling, rather than reducing the price of high technology, as often as not costs, and cost escalation, were aggravated by collaboration itself. [18]

Nau's own work showed that European nuclear collaboration, to say the least, had mixed results, with national interests tending to overwhelm European considerations. Cooperation in aerospace, though more advanced, had an equally equivocal history. In some areas, such as computers and electronics generally, very little progress at all was made in mounting collaborative ventures. By the mid-1970s, with one or two crucial exceptions, technological cooperation in Europe had lost much of its initial momentum. There was certainly little sign of any distinct movement towards greater institutionalised cooperation at a European level, still less of cross-national mergers in high technology industries. In a number of important technological areas, some European states preferred to cooperate with the United States rather than to accept an unpalatable degree of dependence upon their neighbours. [19]

By the late 1970s, a far more complicated pattern of technological and industrial relations between European states, the US, and other members of the western industrialised system, was emerging. It was clear that much of the European fear of a 'technological helotry' had been exaggerated, and that some European products and industries could more than hold their own in world markets. In the case of civil aircraft, this was largely based on collaborative ventures, but many others, even in civil aerospace, were not. More important, Japan had emerged as an equally aggressive and potentially more technologically-advanced challenger to both American and European industry. Nau had predicted that the growing pressure of economic and trade rivalries which he identified in the early 1970s, were likely to increase tension within the western system by the end of the decade. But these trends were, in the event, intensified by American apprehensions about its place in the international political economy, and by the United States' growing vulnerability to the government supported high-technology industries of Europe and Japan.

From the perspective of the mid-1980s, it is increasingly apparent that although there is still a general consensus about the desirability of Western security interdependence, continuing disputes in the economic, industrial and technological fields, have begun to raise serious doubts about the long-term stability of the alliance. In the decade since Nau's book was published, there have been a number of bitter squabbles between the US and its allies on many technological and technologically related subjects. These have included problems in the detailed implementation of the 'common defence', especially in the fulfilment of 'two-way street' philosophies of defence procurement; the balance of trade in all manner of high and low technology products; industrial and technological relations with the Soviet Union; as well as the re-emergence of 'technology gap' fears, this time as a multi-faceted exercise involving the US, Europe and Japan, each with their own particular technological and industrial anxieties.

Some Americans have begun to argue that the United States is losing, or in some respects, has already lost, its long-standing comparative advantage in a number of the strategic, technology intensive industries. To a degree, this has been a 'natural' consequence of broader changes in the nature and operation of the international political economy, and has the general effect of increasing the sensitivity of American politicians to a deteriorating balance of payments and highlighting the stresses caused by deficit budgeting. Most painful of all, especially to legislators facing worried constituents, it has been accompanied by a growing perception that free trade destroys American jobs. In Zysman's view, the United States has become 'more like a normal power', forced into negotiations about the international economic order, with a familiar concern for sectoral interests prompted, as often as not, by domestic lobbies and their Congressional agents. [20] More strident voices see the hand of sinister and deliberate plots to threaten American security. As one US Secretary of Commerce put it, 'foreign economies are taking dead aim on US technological leadership'. [21] Hitherto, any disruption to the free market was localised and the concern of individual companies, which could safely be ignored by American governments. Some economists also believed that a reduction in costs from cheaper imports would benefit the American consumer. From the early 1970s, however, in sector after sector, the United States seemed to be losing control of key markets to foreign companies often supported by their governments. In the view of one tract, 'a world wide misallocation of resources is being accelerated in directions and to a degree that threatens US national interests'. [22]

Zysman for one, feels that the United States Government should take positive and affirmative action to arrest the long-term decline in America's manufacturing base. He challenges the oft-stated American view that government could not, and should not inter- vene in the market. The danger was not simply a distortion of the 'workings of otherwise efficient markets', but a permanent altera- tion in the terms of international competitiveness, leading to an 'irrevocable change in the structure of the market'. Government intervention, especially in those industries which had 'learning curve economies', could play 'an important role in stimulating or hindering (these economies) in domestic firms, and hence in affect- ing the competitive advantage of these firms in international markets'. [23] It should be said that some of the 'threatened' indus- tries, and the American civil aircraft industry is a prime example, do not necessarily accept that direct government intervention is desirable; at least, that is, for the moment. However, there are demands for some kind of action from the United States Govern- ment to counter European subsidies, even if the instruments through which this might be effected are limited or considered to be philosophically doubtful.

America's erstwhile allies and contemporary economic competi- tors naturally see the situation rather differently. In their view, the 'free market', so beloved of American economic liberals, was regarded as the norm precisely because of its intrinsic benefit to the United States. To many Europeans, the United States even now possesses enormous structural advantages in key technological sectors simply because of its size and financial capacity, and as a consequence of the driving technological dynamo of a super- power's defence commitments. In certain respects, and in some areas, notably in commercial aerospace, consumer electronics and in the crisis-ridden field of nuclear power, the United States has lost a commanding role in international markets. But even in these areas, the United States is still pre-dominant, with European and Japanese competitors some distance from achieving equality with American producers. The United States, as the effects of its technology export restrictions attest, can still exert considerable pressure on its colleagues in the Western alliance through its control of key technologies. By the same token, many Europeans are quick to suspect American motives when, as in the case of the Strategic Defence Initiative, the US offers to share technology with its allies. One interpretation is that the US wants European ideas and basic research, but will claim the lion's share of industrial applications. Such fears underpin the French 'Eureka' initiative designed to match US led research into 'Star Wars' technology which migh have far reaching consequences for several vital commercial areas. In short, the international politics of technology increasingly seem to support the old adage, 'where you stand depends upon where you sit'. The view of what is wrong, or what is right about the pattern of technological development, the role of government in promoting industrial development, and the legitimacy or otherwise of claims to a 'rightful share' of world markets in high-technology products, varies according to whether you sit in Washington and Seattle, or Paris and Toulouse.

Yet despite a seemingly irresistable tide of inter-regional competition, there are still significant ties of cooperation and interdependence linking the western industrial states. While some regionally-based cooperative programmes tend to reinforce the pressure of competition and serve to increase the tension between the US, Europe and Japan, other examples are helping to create a framework of inter-regional collaboration. The existence of such ties qualify what might otherwise be a bleak picture of increasing alliance crisis and division. By examining closely the patterns of cooperation and competition in the particularly sensitive area of civil aerospace, we will be in a better position to assess the strength of these parallel trends, and to consider the options facing government in an increasingly complex environment.

The pattern of cooperation and competition: an overview

Civil aerospace products range from small and inexpensive light aircraft and engines, to some of the most complex, advanced and costly examples of high technology. This book is mainly concerned with large commercial aircraft (over 100 seats), and the most powerful engines (more than 10 000 lb of thrust). There are five main centres of civil production outside the Communist Bloc, the United States, Britain, France, West Germany, and Holland. Japan is rapidly acquiring an important civil aerospace capability, and other countries have significant subcontracting industries. Only two companies, Boeing and McDonnel Douglas (MDD) and the European consortium Airbus Industrie (comprising British Aerospace, Aerospatiale of France, MBB of West Germany, and the Spanish company CASA), produce aircraft of over 150 seats. Equally, there are only three companies, General Electric (GE) and Pratt and Whitney (P & W) of the United States and Britain's Rolls Royce, which are capable of developing a large civil aero-engine. The fourth placed manufacturer of civil engines, the French firm Snecma, has based its successful entry into the market almost entirely on collaboration.

Commercial aerospace, then, represents one high technology where European manufacturers have begun to make inroads into an area of long standing American domination. The most public face of this competition is the struggle between Airbus Industrie and the two American giants, MDD and, especially, Boeing. The Airbus also has intrinsic significance as an example of European industrial and technological cooperation. In a region where the rhetoric of joint ventures has often out-run concrete results, the Airbus has become a working symbol of European cooperation. The Airbus programme, and its international holding company, Airbus Industrie, has achieved a degree of industrial and commercial credibility unsurpassed by any other European collaborative project. Although Airbus Industrie is still far short of being a fully integrated transnational company, it has welded much of European civil aerospace into a comprehensive and sophisticated cooperative structure. Although there are European civil aircraft being built outside Airbus Industrie, even by some of its members, the Airbus is the foundation of Europe's challenge to the United States.

Although most attention and political heat has been focused on

Airbus' battle with the American majors, and the possible effect this could have on intra-alliance trading relations, other aspects of civil aerospace have a different pattern. The civil aero-engine industry in particular, is characterised by a much more complex structure of cooperation and competition which cuts across regional divisions. There are, for example, now two international consortia based on trans-Atlantic axes, and a profit sharing arrangement between Rolls Royce and GE; American and European engines fight for places on each others' aircraft; and competitors in one engine sector can be collaborators in others. Even the much-emphasised battle between Airbus and the Americans hides a more complex reality. US airframe companies have established sub-contract and risk-sharing agreements with both European and Japanese firms. The Italians in particular, have more links with American firms in civil projects than with their European neighbours, and the Airbus itself has a substantial American input. Nevertheless, the success of Airbus has led some Americans to demand action from the Federal Government to stave off what is seen as a direct threat to American industrial interests. Equally, some Europeans, notably the French, tend to view Airbus as a symbol not only of European technological unity, but as a focus of resistance to American industrial domination. The existence of at least two distinct patterns of industrial and technological relationships inevitably complicates the analysis of the civil aerospace industry, but perhaps more importantly, it also implies a hidden cost for those who might be tempted to adopt protectionist strategies in response to perceived technological threats.

The core of this book is an exploration of the patterns of international cooperation and competition and their implications for the Western industrial system. Its approach and structure has been largely determined by the history and organisation of the civil aircraft industry since the Second World War. The bewildering array of interrelated commercial, technological and political events which has shaped, and continues to affect civil aerospace is best understood when considered in sequence, a viewpoint which is reinforced by the inherently developmental nature of the technology itself. However, this book is not meant to be a detailed history of civil aerospace since the war, but is an examination of an industry during a period of change. The emphasis is on the events and issues of the last fifteen years, when collaboration has emerged as a necessary answer to fundamental problems of rising costs and increasingly fierce competition.

Chapter 1 outlines in general terms the nature of the commercial aerospace industry, its post-war evolution, and the main centres of civil aerospace production. Chapter 2 is devoted to an analysis of the Airbus programme, and the formation and operation of Airbus Industrie. Chapter 3, in contrast, examines the various American attempts at collaboration with European and Japanese companies in the 1970s and early 1980s. Chapter 4 looks at the engine sector, where a different pattern of relations between the main companies has emerged. Chapter 5 considers the relationship between civil aerospace and the state. There is also a brief survey of current trends shaping the future development of civil aerospace. A concluding chapter both evaluates the experience of collaboration in civil aerospace, and returns to the broader issues of competition and cooperation raised in the introduction.

Notes

1. T. J. Bacher, *The Economics of the Commercial Aircraft Industry*, Boeing Commercial Aeroplane Company, Feb. 1984, p.1.
2. J. Newhouse, *The Sporty Game*, New York, 1982, p.3.
3. Chase Econometric Studies, cited by *The Economist*, 29 September 1985, p.75.
4. Bacher, op. cit., p.1.
5. Newhouse, op. cit., p.6.
6. There is some debate about the exact definition of 'technology', the 'systematic knowledge of industrial arts' according to the Oxford English Dictionary. In modern usage, the term can include 'social technologies' such as systems analysis and other process orientated activity, including 'information technology'. The boundary between science per se and technology has become even more questionable with the arrival of bio-technology and genetic engineering. For the most part, the 'technology' in this book refers to the more traditional arena of industrial hardware.
7. R. Williams, in S. Strange (ed.), *Paths to International Political Economy*, London, 1984, pp.70–1.
8. Ibid., pp.71–83.
9. W. H. McNeil, *The Pursuit of Power*, Oxford, 1983, p.9.
10. See M. Pearton, *The Knowledgeable State*, London, 1982, and M. Kaldor, *The Baroque Arsenal*, London, 1982.

11. R. R. Nelson, *High-Technology Policies*, Washington, 1984, p.1.
12. See, for example, J. Jewkes, 'Government and High Technology', in G. Boyle, D. Elliot and R. Roy, *The Politics of Technology*, London, 1977; K. Pavitt, 'The Choice of Targets and Instruments for Government Support of Scientific Research' in A. Whiting (ed.), *The Economics of Industrial Subsidies*, London, 1976.
13. J. Zysman, *Governments, Markets and Growth*, Ithaca & London, 1983, p.37.
14. H. Nau, *International Politics and International Technology*, Johns Hopkins University Press, 1974, p.13.
15. Ibid., p.12.
16. See, for example, the submission made by the French Trade Union, CFDT, to the French 7th Plan for Aerospace. 'It is the aim of American capitalism to neutralise those elements that are obstructing its imperial will', Annexe 8, p.117, Commissariat General du Plan, Preparation du 7e Plan, *Industrie Aérospatiale*, La Documentation Française, Paris, 1976.
17. J. G. Ruggie, 'International Responses to Technology', *International Organisation*, Summer, 1975, p.566.
18. For further analysis of European collaboration and technology policy in the 1960s and 1970s, See R. Williams, *European Technology*, London, 1973.
19. For example, Snecma and GE, and the European consortium which built the American F16 in preference to either the Tornado or the Mirage 4.
20. Zysman, op. cit., p.51.
21. Cited in P. Harland-Thunberg & M. H. Crawford, *Government Support for Exports*, Lexington, 1982, p.27.
22. Ibid., p.43.
23. J. Zysman & L. Tyson (eds.), *American Industry in International Competition*, Ithaca & London, 1983, pp.24–59.

1 Technology, the Market and the Producers

Technology and the market

The evolution of commercial aviation has been heavily influenced by its ability to 'borrow' technology. [1] In the first instance, this is a product of the direct relationship between civil and military aerospace. Many aspects of early aircraft design were common to both applications. This was particularly so for aero-engines; the jet engine, the most important technological innovation in civil aerospace since 1945, was largely the result of innovation stimulated by military requirements. Although the specialised needs of modern commercial and combat aircraft have diverged, there is still a considerable amount of common ground between the two branches of aerospace. Most of the major civil manufacturers have received at least indirect assistance from active defence programmes. [2] Civil aerospace has also borrowed extensively from technological improvements in other areas, particularly from the metallurgical, materials, petro-chemical and electronic industries.

A result of this cross-fertilisation has been that the civil aerospace industry has developed products of increasing complexity. Consequently, the industry has had to rely on a growing number of component and equipment companies. The industry now consists of a small number of prime airframe and engine manufacturers, supported by an array of dependent subcontractors and suppliers. [3] Greater complexity has brought higher levels of technological uncertainty, has increased the lead time of development, and has had a dramatic impact on development costs. The non-recurring costs of a new aircraft or engine are now in the region of $1.5 billion, and a company's total investment could exceed $3 billion. To recover in full these costs, as well as the interest on loan capital, a manufacturer needs to sell between 350 and 400 aircraft in the first ten years of production. Historically, few projects have achieved this target. Under these circumstances, firms are under considerable pressure to increase the volume of sales. This enables them to extract the full value of the 'learning curve' phenomenon, where costs diminish in a direct relationship to numbers produced. Where possible, they also seek to modify existing types through modest, low cost technical improvements. This is best facilitated through the creation of a 'family' of related aircraft or engines, where existing components and design features can be incorporated into a succession of products. This not only reduces development and production costs, but offers the customer 'commonality' across a fleet of airliners. Finally, the pressure of increasing costs has led many engine and airframe prime-contractors to adopt collaborative and risk-sharing strategies. In some cases, it has also implied a greater role for government in the financing of civil aerospace.

Figure 1 The Cyclical Nature of the Aircraft Business Increases Manufacturing Risk: annual deliveries—total US and European programmes

Source: Boeing

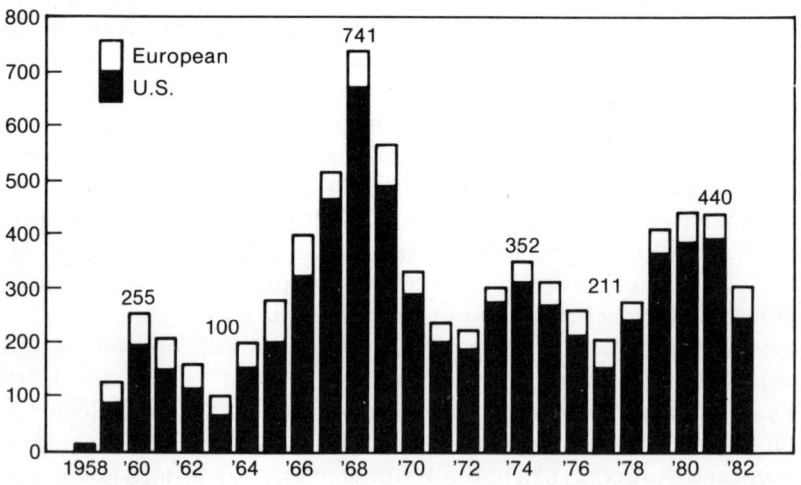

The market for civil aircraft since 1945 has followed a cyclical pattern of high demand and precipitate slump (See Figure 1). This is, in part, a reflection of general movements in the international economy, but it has been made worse by the volatility of the airline business. There is also a direct relationship between technological innovation and demand. New technology has consistently improved airline productivity and competitive pressure has forced airlines continually to invest in new equipment. However, during periods of traffic growth, airlines have tended to overestimate their capacity requirements, and often have then over-reacted at the onset of a recessionary period with deferments, cancellations and a reluctance to order new aircraft. [4]

The immediate post-war years saw both a rapid recovery and an expansion of civil aviation. In the United States a vast new domestic market for air travel appeared, and American manufacturers, with readily convertible military transports, were more than able to satisfy demand. American domination quickly extended to the emerging network of international long-haul routes. In Europe, most airlines faced difficult problems of re-construction, and usually turned to the US for aircraft. The 1950s, with the introduction of the jet airliner, was the most critical phase in the post-war evolution of the market. The jet engine ultimately enabled the construction of faster, heavier and more productive airliners. The British were the first to use the jet in civil aircraft, taking advantage of an initial lead in jet engines to launch the Viscount and Comet. The Americans were slower to use jets for two main reasons. As the established market leaders, US manufacturers were unwilling to embark upon a new, high-cost, high-risk technology until the advantages of the jet had been proven. They also waited until more powerful and efficient engines were available to build a genuinely economic airliner. [5]

In 1952, Panam's order for Comet 3s galvanised the American aircraft industry into action. Boeing launched the Type 367-80, which formed the basis of both the military KC135 and the 707 airliner. Douglas quickly followed suit with the DC8. These two long-haul jet aircraft triggered a 'rush to jets' by the world's airlines which rapidly spread to short and medium haul routes. Britain's lead evaporated through a combination of accident and unfortunate commercial and technical judgements made by British manufacturers and their domestic airline customers. [6] By the end of the 1960s, with the Americans first to exploit the even greater power and efficiency of the large turbo-fan engine, the US had established a monopoly in the field of long-range airliners and a stranglehold over many of the remaining market sectors.

If the introduction of the long-range jet led to a rapid expansion in airline business, it was quickly followed by a slump in traffic. The fall in airline revenues was aggravated by the collapse of the second-hand market, now consisting of virtually unsaleable piston-engined aircraft. The next up-turn in the mid-1960s was encouraged by the availability of economic and highly popular short and medium-range jets. The result was a record improvement in airline revenues, and the manufacturers reaped the benefit of a new wave of orders. By 1969, the 'inevitable crisis' returned, and airlines again found themselves with an excess of capacity, as well as a high level of debt incurred during the conversion of their medium- and short-haul fleets to jets. [7] Demand for new equipment once more fell, and this time, the imminence of technological change did not immediately generate a new market surge. The large, high bypass turbo-fan engine allowed the construction of bigger and still more efficient airliners. Yet the wide-bodied airliner, offering improvements in seat-mile costs of up to a third better than existing aircraft, as well as the prospect of supersonic airliners, failed in the short term to affect airline procurement. Indeed, Boeing introduced the 747 too early, and was nearly bankrupted by the combination of high development costs and poor initial sales. Other wide-bodied projects, the MDD DC10 and the Lockheed L1011, had an equally hard time, made worse by direct competition in a depressed market. The SST was undermined by environmental as well as economic doubts; the US SST was cancelled and the Anglo-French Concorde proved to be a commercial disaster.

In the early 1970s, economic, and consequently market uncertainty, was increased by the oil crisis of 1973 and the following period of 'stagflation'. During this period, Lockheed finally decided to abandon civil aerospace to concentrate on the 'safer' defence market. By the end of the decade, with airlines looking for fuel efficient and quieter aircraft to meet increasingly stringent airport noise standards, the market began to recover. However, general recession returned more rapidly than expected, and its effect on the civil aircraft industry was intensified by the financial vulnerability of many airlines. In 1983, the cumulative debt of US airlines was $10 billion, and $9.4 billion for IATA members. Where a 13% return on capital was regarded as a realistic break-even point for profitability, airlines were averaging only 1.7%. As a result, another wave of cancellations and deferments swept through the aircraft industry.

In addition to these objective problems, the airline business had lost a protective mystique within the financial community; investors were at last treating it as they would an 'ordinary', high risk, ailing industry—with great care and ever greater scepticism. According to Dan Krook, Deputy Chairman of Fokker, 'the banks don't believe in airlines anymore'. [8] Worse still for the manufacturers, a run of airline failures led to a glut of new, or nearly new second-hand airliners. Under these circumstances, the airlines tended to look for leasing deals rather than outright purchases. Some regions, such as the Middle East, and some of the better managed airlines elsewhere, escaped the worse of the recession. These became the targets for increasingly intense and bitter sales battles waged by the aircraft and engine manufacturers. In a buyer's market, customers were able to drive hard and very favourable bargains, further depressing company cash flow.

In the mid-1980s, the market again began to improve with the general upturn in the world economy. The depth of the recession, and the long-term effect of high debt ratios was such that the lag between general economic recovery and the upturn in airline profitability was longer than in the past. Falling fuel prices and predictions of relatively stable fuel costs through to the end of the decade, also had an impact on the market for new aircraft. The advantage of fuel savings bought at a high capital cost was no longer so obvious, and the pressure to sell-off older, less fuel efficient aircraft was not so intense. The total 'life-time' cost of new aircraft, rather than the savings offered by costly technological innovation, became the main criterion for airline judgement. As one airline executive observed, 'one of the hardest things for us to get across to manufacturers is that what we are prepared to pay bears no necessary relationship to their cost of engineering and manufacture. We are simply unwilling to pay so much for an aircraft if we have no hope of earning a reasonable return'. [9]

Nevertheless, the potential market for replacing obsolescent aircraft, especially in the small- to medium-sized category, is huge.

Most estimates indicate that the market for new aircraft by the end of the century will be worth between $150 and $200 billion. [10] 1985 may well have seen the start of this process; with 654 aircraft ordered, worth close to $23 billion, it was the second best year ever for aircraft sales. However, despite these positive signs, airline finances were still fragile, and uncertainty about the exact timing of the new technology, again encouraged some airlines to delay firm commitments until the position was clearer. The cyclical nature of the market may be less pronounced in the future. By the 1990s, airline expansion should begin to match the general growth in gross national products. Sudden changes in the business environment, such as a sharp rise in fuel costs, which could disturb an orderly pattern of airline procurement, are expected to be less likely. This could have the affect of removing the extremes of uncertainty experienced since the war. [11] Even if the market becomes rather more predictable, it will not eliminate the risks faced by manufacturers. Timing the launch of a new airliner, either to catch or to stimulate a new cycle, has been, and will remain critical. As Phillips puts it, 'too soon or too late, too radically or too conservatively—each may be a source of failure in the market. The proper balance is so difficult to achieve that the market, by rewarding the successful and punishing the failures, tends to be concentrated'. [12]

The changing commercial environment

American manufacturers, and Boeing in particular, still dominate the market for civil aircraft and engines (See Table 1). Up to the mid-1970s, the US had 96% of the market for civil aircraft in the non-Communist world; in 1980, the US still provided 86% of the world's airline fleet. In the long haul sector, this was as high as 99.1%. American penetration of Europe was 76.7%: put another way, Europe was importing 2.5 times by volume as much US made civil equipment (including American components used in the Airbus) as it exported. [13] The Airbus and other smaller European jet airliners have begun to make an impact, and in 1985 European manufacturers held 38% of outstanding orders. The position has also been better in the case of civil engines, where Rolls Royce historically has had a strong presence in the market, supplying half of the world's civil turbine engines up to 1965. [14] Rolls has

Table 1 Total World Commercial Jet Transport Aircraft Deliveries, 1958–85‡

	Deliveries		(On order‡) excl. options	
707*	957	(last commercial delivery 1979)		
727	1831	(last delivery 1984)		
737–200†	1086		31	
737–300	53		215	
747†	617		30	
747–400	—		10§	
757	49		13	
767	121		72	
DC8	556	(last delivery 1972)		
DC9†	976	(last delivery 1985)		
MD80	237		157	
DC10*	369		8	
L1011‡	249	(last delivery 1984)		
880/990	102	(last delivery 1965)		
US sub-total	7203	(83.7%)	371	(61.9%)
Comet*·†	112	(last delivery 1967)		
Caravelle	279	(last delivery 1973)		
Trident	117	(last delivery 1978)		
VC10†	54	(last delivery 1970)		
BAC-111	232	(including Romanian licence production)		
F28	222		15	
F100	—		38	
Mercure	10	(last delivery 1975)		
A300	253		17	
A310	71		46	
A320	—		90	
BAe 146	30		23	
VFW 614	10	(last delivery 1977)		
Concorde	14	(last delivery 1980)		
Eur. total	1404	(16.3%)	229	(38.1%)
Total	8607	(100%)	600	(100%)

Sources: Boeing Commercial Airplane Company; Flight International.
* Military version produced (707–KC135, AWACs – over 904 delivered/ordered; DC10–KC10 – 60 delivered/ordered; Comet–Nimrod – 48 delivered/ordered)
† Military orders
‡ As of October 1985
§ Launched November 1985

maintained a strong third position in the market, and Snecma's collaboration with GE has further helped to reduce the imbalance.

A number of factors have contributed to Europe's poor commercial record. Even where narrowly drawn specifications did not hinder European contenders, superior marketing, customer liaison and support tended to give American firms an important edge. The sheer size of American firms, combined with the assurance of large initial orders from their domestic customers, led to greater efficiency, better delivery times, and a greater sensitivity to market requirements. US firms also derived an advantage from higher productivity, though this was offset to a degree by cheaper European labour costs. [15] The huge, integrated American domestic market was perhaps the single most decisive factor in explaining American success. Orders from two of the larger US airlines could easily finance the launch of a new aircraft. Put another way, if a US firm could win 25% of its home market, a project rapidly moved down its learning curve, 'but for the (European) manufacturer, even 100% of the home market still leaves him with cost problems relative to the US manufacturer...'. [16] Moreover, while a good foreign product could make headway in the American market, tariffs, political and financial pressure (especially from the banks that financed both airlines and manufacturers) and 'natural' buy American feelings, conspired to obstruct European sales. European companies, with some justice, felt that the US Government was better at supporting its industry through export credits and diplomatic inducements, than were their own. [17]

On the other hand, European companies have received direct aid from the state. This has been justified as providing compensation for the smaller financial base of European manufacturers, and to match the indirect help of American defence spending. Although commercial objectives were important, support for many otherwise uneconomic projects was given to promote employment, technology and balance of payments policies. The growing success of the Airbus in particular, has increased the political sensitivity of the subsidy issue. This is not new; both the Comet and the Concorde triggered comparable fears of an impending loss of American leadership and similar demands for action to counter European subsidies. This time, however, there is a much stronger feeling that European competition is serious and permanent, and that American industry cannot count upon command of the commercial high ground indefinitely.

US manufacturers have been hindered by changes in their 'home market', particularly those brought by domestic airline de-regulation. De-regulation has had a profoundly de-stabilising effect on the US airline industry. A number of airlines have collapsed, and while new airlines have sprung up, they are not necessarily committed to American equipment. The tendency is also towards the purchase of smaller aircraft with lower margins for the producer. Although four airlines still have 56% of the home market, even they are more reluctant to spend large sums launching new aircraft. Frank Bradley of Citibank summarised the growing problem for US manufacturers, 'although the few strong carriers are possible launching platforms for new aircraft, the broad platform which the US industry has provided in the past, and upon which US manufacturers built their commercial business, will not be there'. [18]

Other trends in the commercial environment have also helped the Europeans to compete more effectively with the Americans. The expansion of the 'Rest of the World' market, especially in the Middle and Far East, where regional factors tended to insulate countries from the recessionary effects of the 1970s and early 1980s, has been particularly important. Growth in Asia and the Pacific Basin is likely to be very strong, with higher than average increases in national GNPs. This, combined with a geography well suited to air travel, should ensure a promising future for sales in this area. [19] These airlines are not only less committed to established suppliers, but are also more likely to be swayed by financial and other political inducements. Nevertheless, despite these changes, Europeans believe that US companies still have a considerable head start over foreign competitors because of their domestic market. Between 1985 and 1993, IATA estimates that 40% of orders by its members for new aircraft will be placed by American carriers. Overall, American airlines are expected to fill over 54% of all new orders up to 1999. Even Airbus has only sold to three American airlines since 1976, and further success in the American market is regarded as being important, if not vital, to the future health of the programme.

An airline's decision to buy a particular aircraft or engine is based on a number of related factors. There are certain fundamental economic criteria; range and payload, combined with capacity and frequency demands; airport and environmental requirements; the capital costs of acquiring the aircraft, training crews and maintenance; and, especially important since 1973, fuel efficiency. Similarly, fleet standardisation and commonality within a single airline, or between a number of airlines 'pooling' equipment and ancillary services, influence an airline's decision. [20] Where there are alternative engines, the struggle between manufacturers follows its own competitive path. Although technical and economic merit is a primary consideration in airline procurement, the availability of a favourable financial package, always useful on the margin, has become very important in winning orders. By the same token, the existence of a national presence in a product plays an increasing part in airline choices. This has been particularly so in Europe, where nationalised carriers have encountered enormous, and often irresistable pressure from governments sponsoring their indigenous aircraft industry. Beyond this, other political factors, totally unrelated to the airline and aircraft industries, can influence an order.

To summarise; up to the early 1970s, civil aerospace has been characterised by the dominance of the US, qualified only by spasmodic technical and commercial challenges from European manufacturers. This has been explained by a variety of American advantages and European inadequacies. The most important aspect of the last twenty years, however, is that, despite some very expensive setbacks, the European industry has survived, and is now in a position to play a major role in the civil aerospace market. Indeed, either as a result of intra-European or trans-Atlantic collaboration, European industry is better placed now than at any other time since 1945, to effect a strategic shift in the commercial balance of power in civil aerospace.

The Producers

An international industry

Technological and commercial pressure has reduced the number of prime contractors in the international civil aerospace industry. In the case of European companies, the consolidation of national manufacturers and the adoption of cooperative strategies, has been to some extent a response to American competition. But as the ranks of prime contractors have diminished, the number and national spread of subcontractors and risk-sharing partners has grown. This has been caused by deliberate efforts on the part of the prime contractors to spread the burden of development and production, or is the result of offset demands from purchasing states. For example, as a result of a deal between MDD and the Chinese People's Republic, MDD will finish the first 25 of 50 MD80s bought by the Chinese national airline, the remainder will be built by the Chinese. In the US alone, the proportion of subcontract work has risen from a typical level of 40% for the Lockheed Electra in the 1950s, to over 70% for the Boeing 747. In the 1960s, the Americans began to extend subcontracting to foreign firms; for example, participants in current Boeing programmes include Canadian, Japanese, Italian and British firms. A similar pattern obtains for Airbus Industrie and other European projects. Significantly, many of these companies are now risk-sharing subcontractors, carrying the cost of developing their part of the main programme in return for a share of the profit. [21] There are, however, six countries, the United States, Great Britain, France, West Germany, Holland and, increasingly, Japan, which constitute the heart of the international civil aerospace industry.

The United States

Currently, with over 60% of the global market in 1985, Boeing dominates both the American and the world civil aircraft industry. Although it is also a major defence contractor and has increased the value of its military work, more than half of its business is in the civil sector. Boeing is also one of the USA's leading exporters, with over half of its products going overseas. Boeing emerged as the dominant force in international civil aerospace with the advent of the jet airliner and the success of the 707. The 707 was followed by the hugely successful 727 and 737 airliners. However, the cost of developing the wide-bodied 747 nearly bankrupted the company. At one point, Boeing's creditors held a syndicated debt of $83 million. Collapse was averted by ruthless cuts in manpower and a substantial increase in productivity. Boeing also adopted a more conservative development strategy, emphasising the 'family' approach to new projects. Boeing recovered to win 43% of the wide-bodied market, and a near monopoly of long haul sales. During the 1970s, the company launched two new aircraft, the 757 and 767, an investment of over $3 billion, more than double the company's net worth. Boeing also launched the 737-300, as an interim contestant in the crucial 150 seat, '727 replacement', airliner battle. More recently, Boeing announced the launch of a larger version of the 747, and is developing a new aircraft utilising the revolutionary prop-fan, or UDF engine. Boeing has recovered strongly from the recession of the early 1980s, and 1985 saw its second best year in terms of sales (390) and a record year by value ($15 billion). Overall, even by the high standards of the American aircraft industry, Boeing has a superb reputation for efficient production, marketing and product support, matched by an aggressive, even arrogant reputation in its dealings with potential collaborators. [22]

The second ranked civil aircraft manufacturer is McDonnell Douglas (MDD), a product of the 1966 merger between Douglas, one of the truly legendary names in civil aviation, and the defence contractor McDonnell. During the 1960s, MDD looked as though it could regain ground lost during the jet revolution to Boeing with its wide-bodied DC10. However, under pressure from the Lockheed L1011 and the European Airbus, and following a series of accidents, the DC10 never fully capitalised on its early promise. During the 1970s, MDD was able to launch a derivative 150 seater, the DC9-80 (MD80), but with unprofitable civil programmes (one estimate put MDD's cumulative loss on civil aerospace at $450 million) undermining the company's overall profitability, the future of its civil operation was uncertain. In November 1983, production of the civil DC10 and development of many other long-term projects were terminated. Only the MD80 was left to carry on the Douglas tradition. The KC10, a military tanker version of the DC10 also remained in production. [23] At this point, it was tempting to write-off MDD as a major civil producer. However, an aggressive campaign was launched to sell the MD80, and led to what was at the time, the largest single order in airline history. In June 1984, MDD restarted production of the civil DC10, and is currently considering the launch of a updated version, the MD11. Like Boeing, MDD is also investing heavily in a prop-fan airliner. In 1984, MDD won 43% of the year's sales, leading James Worsham, President of Douglas Division, to observe that 'we are most certainly not out of the commercial business'. But despite this vigorous come-back, MDD's commitment to civil aerospace still remains somewhat equivocal. [24]

The US civil engine industry consists of two companies, Pratt & Whitney (P & W) and General Electric (GE). Both have extensive defence interests and are part of larger industrial conglomerates. GE developed America's first jet engines during the Second World War, but in the immediate post-war years concentrated on military work. GE's initial venture into civil jets was seriously damaged by its association with the commercially disastrous Convair 880/990 airliner. [25] P & W, on the other hand, took to the civil market, developing the JT3 for the 707 and the DC8. By 1968, P & W had captured 90% of the civil market, and had reinforced its dominance by developing a large turbo-fan engine, the JT9, for the 747. The JT9 was not without its problems, and there was a growing feeling in the aircraft industry that Pratt's dominance was unhealthy. With Rolls Royce equally *hors de combat* for the early part of the 1970s, GE had an opportunity to re-enter the civil market. [26]

GE's breakthrough was helped by building the CF6 for the C5A military transport. The CF6 was particularly well suited to the L1011, DC10 and the Airbus. Although it lost the L1011 to Rolls, the early commercial penetration of the DC10 and its monopoly of the Airbus, gave GE a vital edge over both P & W and Rolls Royce. [27] The success of the CF6 generated cash flow for derivatives which have been used on the 767 and the A310. GE's links with Snecma, founded on a joint CF6 programme for the Airbus, led to the CFM56, which enabled it to broaden its engine range at a reduced cost. The result has been a very impressive growth in GE's share of the market. (Table 2) P & W's response was a $700 million programme to develop the P & W JT10 (PW2037), and an aggressive sales campaign to regain its position. P & W also looked to collaboration as a way of reducing the financial burden of maintaining a range of engines; in 1983, P & W joined Rolls, and five other companies, to produce the V2500 to compete with the CFM56. Both P & W and GE have active prop-fan research programmes.

American companies claim that they get little or no aid from the state for civil aerospace. Indeed, the general belief is that public subsidisation of civil aerospace undermines the effectiveness of the market as a mechanism for winnowing out commercially and technically weak products. [28] Yet the development of American

Table 2   Share of market for civil aero-engines

	1966 (%)	1978 (%)
P & W	92.4	62.7
Rolls	5.9	12.7
GE	1.7	24.6

civil aerospace is, or has been, facilitated by less overt forms of state intervention. This has included technological spin-off from military and space research; the financial links between military and civil programmes; the 'GOCO' (Government Owned, Contractor Operated) arrangement for plant and equipment; favorable depreciation allowances and other tax credits to US airlines which have been used to buy American aircraft; and support for overseas sales through the Eximbank. [29]

Since the Second World War, the United States has possessed the world's strongest and most successful civil aerospace industry. Nevertheless, American firms have not been immune to the effects of rising costs and market pressures. The increasing presence of the Airbus in the market has certainly worried Boeing and MDD, and was one of the reasons for Lockheed's decision to end civil production. Its development represented a new and uncertain factor in a market hitherto largely shaped by the actions of an American oligopoly. The Airbus is viewed as deliberate strategy on the part of the Europeans to attack the civil aircraft business through subsidised production. Although Boeing in particular has called for US Government action against 'unfair competition', the nature of any response is yet to be defined. There is no doubt, however, that further inroads into American sales will generate more pressure for counter-measures, either through tariffs, or, more likely, by an enhanced Federal commitment to support civil aerospace research and development.

American industry has recognised that international collaboration can be a way of easing its financing problems. This has had both positive and negative aspects. On the one hand, as John Brizendine of MDD put it, 'we ought not go down separate roads, spending the money twice instead of once. There's enough to share—and Lord knows, there's enough risk to share'. [30] On the other, American companies have been concerned to get inside any European Community tariff barrier which might have been raised against non-European products. Transatlantic collaboration has also been seen as a possible way to nullify or dilute growing competition from European firms. Anti-trust obstacles to intra-American cooperation, the overstretching of domestic sources of launch capital and a reduction in Federal funding for civil R & D in the late 1960s, were other, more immediate reasons for seeking overseas partners. [31] During the 1970s, all the major American firms tried to reach collaborative agreements with foreign companies. While American engine companies successfully participated in a number of joint ventures, the airframe manufacturers were rather less willing to make the compromises necessary to collaborate with the larger European companies. Nevertheless, every major American civil aerospace company now has a range of partnerships with overseas manufacturers.

Europe

The importance of aerospace in European technology policy has long been recognised. According to one European Commission report, it was 'one of the chief representatives of a type of employment—highly skilled, commanding sophisticated technologies and a high level of investment—toward which the Community must necessarily move in the future as the industrialisation of the Third World proceeds and a wider division of labour unfolds'. [32] Since the late 1950s, with nationally-based industries under increasing pressure, European states have turned to joint ventures as a solution to the problem of launching high cost aerospace projects. Motives for collaboration have inevitably contained a mixture of political and economic considerations. Collaboration was seen as a way to reduce the risks of development and to expand the market base for aerospace products, and thereby to compete more effectively with the Americans. [33] In practice, collaboration has often proved to be more problematic, and its benefits have not always been so direct as at first envisaged. Joint programmes have incurred extra costs as a result of managerial inefficiency and politically-inspired industrial compromises. One estimate puts the difference between a collaborative and a comparable national project at between 10% and 30% on development costs, with an increase of up to 10% on production. There were other problems which could lead to 'substantial transaction costs, delays in development, compromises in operational requirements for military aircraft and possible losses of property rights in national technology'. [34] But collaboration remains the most practical European approach to the construction of large civil airframes, and to a lesser extent, for engines. Over the last twenty years many technical, political and organisational lessons have been learned. Some of the petty sources of dispute and contention which plagued early joint ventures have been eradicated, or have greatly diminished. Yet the chequered history of cooperation has left its mark on European countries. There are still differing national perceptions of the role and function of collaboration, and differences in priorities between European states.

Despite twenty years of active European cooperation, there is no European Community policy for civil aerospace, or for the aerospace sector generally. In March 1975, the European Council passed a resolution calling for 'concerted action and consultation' amongst the member states on aeronautical policy and required the Commission to submit a further report 'on the conditions in the industry and the measures necessary for its development'. This resulted in the most ambitious statement of 'European' policy towards aerospace, the Action Programme for European Aerospace of October 1975—the Spinelli Report. [35] In many respects, Spinelli did little more than repeat a well-known diagnosis of European aerospace ills, and, perhaps inevitably, the Report had a limited practical effect. [36] Although the EC Council accepted the Report's basic arguments, it failed to implement any of its main proposals. The Commission eventually came to the conclusion that it had been 'overtaken by events' and decided not to pursue matters further. [37] On balance, Spinelli probably overplayed his hand, especially in the Report's implicit demand for a substantial increase of Commission authority in aerospace matters. The major European aeronautical powers were happy with the existing compromise between national and regional approaches, and its under-current of anti-Americanism offended those states, notably the Italians, which had profitable links with American companies. [38]

The prospects for a 'European' policy for civil aerospace has always been complicated by differing views about relations with the US. At one extreme, cooperation with American companies has been seen as a 'kiss of death', which would condemn European companies to an insecure future, 'metal bashing' for the Americans. At the other end of the spectrum, the engine industries of both France and Britain have so far preferred to forge links with the US rather than to cooperate with each other. Some of the smaller European industries fear equally domination by their larger neighbours as much as by the United States. Given the choice, many Europeans would prefer a partnership from strength approach; a united European initiative to share costs and risks with one or more American companies. As Martin Grüner, the German Aerospace Co-ordinator, put it, 'a united European aircraft industry can effectively engage the much vaunted transatlantic marriage. Individual negotiations by European companies will weaken rather than strengthen this potential'. [39] In some areas of civil aerospace, notably in trade and tariff negotiations, the European Community has successfully acted on behalf of its member states. Equally, there is possibly room for an EC role in other 'functional' aspects of aerospace. [40] But for the most part, European civil aerospace remains a patch-work of international structures based upon national industries with their own separate identities and priorities. Arthur Reed summed this up nicely, 'in spite of the accelerated trend towards cross-national cooperation...aircraft industries will remain an important economic and industrial arm for their home countries'. [41] The future may yet lie in a more integrated approach, but for the present and the near-term, the most realistic proposition for European cooperation is the project-based consortium like Airbus Industrie.

Britain

Britain has a long tradition of civil aerospace development. Indeed, after the Second World War, British technology frequently led the world. However, for a variety of reasons, British industry was never able fully to capitalise on its evident skills and ability. [42] As a result, British civil aerospace has usually had to depend upon state aid for development, or the support, sometimes double edged, of the nationalised airlines.

The British civil aerospace industry now consists of one main airframe manufacturer, British Aerospace (Bae) and one engine company, Rolls Royce. Shorts produce a range of small civil airliners, as well as subcontracting to Boeing and Fokker. BAe was formed in 1977 with the merger and nationalisation of Hawker Siddeley Aviation and the British Aircraft Corporation. Between 1980 and 1985, BAe was 'privatised' by the Conservative Government. Civil aerospace only represents about 20% of BAe's overall activity, but it does entail a substantial financial risk. In 1982, £100 million was written off on civil development. Similarly, although its profits for 1984 were £81 million on sales of nearly £2500 million, profits on civil aerospace were just £7.5 million on £572 million in sales. A further £51 million was written off on civil development. The 1985 results showed a £2.5 million loss on a civil turnover of £299.6 million. [43] BAe is a major shareholder in Airbus Industrie. It also produces (with Avco of the US as a risk sharing partner) the BAe 146, an 80–100 seat jet airliner, the 748 twin-turbo prop, and has recently launched an Advanced Turbo-Prop feederliner. Although its civil operation has not been profitable, BAe believe that the Airbus, and its other civil projects will be commercially viable, and increasingly an important diversification of and supplement to its military work. Since the mid-1960s, the British airframe industry has been a strong advocate of European cooperation, and during the 1970s, resisted pressure to join American projects.

Rolls Royce is one of the world's leading aerospace companies, and by far the largest engine company in Europe. About 40% of Roll's business is devoted to the RB211 and other civil engines. In 1971, largely as a result of its contract with Lockheed to build the RB211, Rolls was forced into bankruptcy and 'temporary' nationalisation. In 1985, it was announced that Rolls would be privatised before the next General Election, with nearly £500 million of debts likely to be written off by the government. [44] In 1984, Rolls reported its first profit in five years—£326 million on sales of £1403 million, with outstanding orders for over £2.5 billion. An important reason for the company's return to profit was an enormous improvement in productivity (up to 40% in some areas— representing a reduction of 20 000 employees since 1980). Through a mixture of independent and collaborative projects, primarily with P & W, GE and the Japanese, Rolls can offer a range of engines from 5000 lb to over 60 000 lb of thrust. However, as a company almost entirely concerned with aero-engines, Rolls is particularly vulnerable to the cyclical nature of both civil and military markets.

Unlike the airframe industry, in developing civil engines, Rolls has pursued a transatlantic strategy, preferring both US customers and American partners to those in Europe. As one senior Rolls executive put it, whereas the British airframe industry had to collaborate, Rolls Royce could 'compete on all fours with Pratt & Whitney and General Electric'. In the 1970s, Rolls recognised that cooperation in some engine categories was essential, but again its preference was to seek partners outside Europe. Sir Kenneth Keith, Rolls' Chairman in the early 1970s, summed up the difference between cooperating with the Americans and the Europeans in these terms: 'The advantage of our collaboration with Pratt & Whitney is that there is an exchange of technologies between equals; in certain areas they are better than us, and in certain areas we are better than them. Much of our collaboration within Europe has been in a one-way direction, with Rolls Royce on the giving end.' [45] Rolls does have links with the smaller European companies, Fiat and MTU, but largely as a consequence of prior agreements between the Europeans and P & W.

Since the Second World War, British Governments have varied in both the extent and the enthusiasm with which they have supported civil aerospace, but aid in some form has generally been forthcoming. The return on government assistance to civil aerospace has been poor, often non-existent, and aid often had to be justified by reference to employment, technological and balance of payments factors. [46] Support for civil aerospace has generally been given in the form of launch aid covering a proportion of the non-recurring costs of development. Repayment has usually been based upon a levy on sales of the aircraft or engine. Some projects, such as Concorde, have been financed by direct government contract. As nationalised companies, BAe and Rolls Royce have had access to public capitalisation, some of which has been used for civil programmes. Finally, in the past, the nationalised airlines have been expected, or have been compelled, to 'buy British'. In recent years, British Governments, and particularly the Thatcher administration, have stated that civil aerospace must be commercially viable. As one Conservative Industry Minister put it, 'the aerospace business is for making profits, it is not a form of occupational therapy'. [47] Nevertheless, non-commercial factors have continued to be important determinants of civil aerospace policy.

If there has been a long history of government involvement in British civil aerospace, governments have rarely possessed an overall strategy for the industry. The divergence of interest between the airframe and engine sectors has not helped the formulation of consistent policies, but the industry has been unable to rely on successive governments to maintain a stable pattern of support for civil aerospace. The question of ownership has all too often been the main area of interest for British politicians. Although the privatisation of BAe was generally welcomed by management, Sir Raymond Lygo, BAe's Managing Director, warned that privatisation should not put BAe at a disadvantage compared to its European colleagues, so that 'we will be driven by considerations of profit and loss on a short term basis, which may well preclude our participation in long-range projects or indeed collaborative projects with foreign government-supported industries'. More fundamentally, the SBAC, the industry's trade association, has argued that Britain, and British Governments, should adopt a more deliberate strategic approach to aerospace similar to that of the French. According to the SBAC, the French have been 'single-minded' in their 'pursuit of high-technology as an article of faith'. Such unequivocal support for the aircraft industry was responsible for France's ability to 'challenge the UK's position as second to the US in the Western World'. [48] However, despite past problems, the British civil aerospace industry as a whole is a vital element in Europe's bid to match the United States.

France

After the Second World War, the French aircraft industry fought hard to establish an independent presence in the world civil market, but like the British, they found that the financial burden was too great to bear alone. From the 1960s, the French have looked to collaboration as a means of developing the technological and financial strength to take on the Americans. Although France has lacked the all round capability of the British aircraft industry, it has emerged as the leading force in European civil aerospace.

There are two main airframe companies in France; Aerospatiale, responsible for most French civil aircraft projects; and Dassault, mainly involved in military programmes. Aerospatiale is the product of largely state inspired rationalisation, emerging in its present form in 1970, with the merger of Nord and Sud Aviation. Through its constituents, Aerospatiale has been in public hands since the mid-1930s. It is a major shareholder in Airbus Industrie, and has also cooperated with the Italian company Aeritalia to build the ATR42 feeder-liner. Dassault's success has been based on profitable military contracts and overseas arms sales. However, its acquisition of Breguet in 1967, was partly aimed at increasing its capital and production base in order to undertake more civil work. Dassault produced a successful series of business jets, but its only venture into large civil airliners, the Mercure, was a commercial disaster. Attempts to resuscitate the programme in collaboration with MDD were unsuccessful. Dassault is, however, associated with various Dutch civil aircraft, on subcontract to Fokker. Aerospatiale has sometimes been unfavourably compared with Dassault's privately-based success, but both have received similar and substantial support from the French Government. In 1984, Aerospatiale reported a modest profit, helped by a more diversified product range. [49] In the 1980s, the state acquired shares in Dassault, and currently owns 46% of the company.

The nationalised company Snecma is the main centre for French civil aero-engine development. Historically, the civil engine sector was one of the weakest aspects of French aerospace. Mergers again helped to strengthen Snecma's overall position, and the company benefited from France's ambitious military goals. The decision to upgrade its civil capability was a deliberate and bold step. Snecma recognised that collaboration was the only realistic way this could be achieved. Snecma had considerable experience of cooperation, both with the British firm Bristol Siddely and with P & W. Involvement in the Anglo-French Olympus engine developed for the Concorde, brought a considerable improvement in Snecma's technical competence, but cooperation with Rolls Royce to build an engine for the Airbus was not so happy. Neither company, especially amongst senior management, was warm towards cooperation, but the heart of the matter was French unwillingness to accept a subordinate status to Rolls Royce within a European industrial context. When Rolls withdrew from the Airbus, Snecma joined GE in producing the CF6 for the A300. This was followed by a more comprehensive joint programme to develop the CFM56. As Christopher Layton put it, cooperation with GE was with a 'conveniently distant friend, whose power and technology might help Snecma to build up a position nearer equality with the British'. [50] Snecma has no ambition to become an independent producer of large civil engines, but in just over a decade, it has achieved a well deserved fourth place in the world civil engine industry.

To the French, aerospace has been closely associated with wider national political objectives. As a 1977 Parliamentary report noted, 'more than any other sphere of activity, aerospace is a test of strength between states in which each participant deploys his technical and political forces'. [51] This was consistent with a wider policy of national independence in key technologies—a policy which reached its zenith under De Gaulle, but which has been followed generally by French Governments since 1945. The state naturally assumes a central role in all aspects of aerospace. According to one Prime Minister, it is 'an omnipresent actor', taking decisions designed to ensure the cohesion of French activities and the enhancement of French capabilities. [52] During the 1960s, a more overt politicisation of French civil aerospace became apparent. The defined aim was to challenge American domination of the civil aircraft industry, and to utilise the industrial and technological benefits of civil aerospace in revitalising the French economy. It should already be clear, though, that such sentiments have not prevented the French from forging links with American firms when it has suited their interests. The growing importance of civil aerospace was recognised by the Sixth Plan of 1970, with substantial budgetary allocations in support of the Concorde, Airbus and Mercure. Direct contracting was replaced by *advancés réimbursables*, a method of repayable aid to cover non-recurring costs similar to the British system of launch aid. Assistance has also come in the form of industrial credits and borrowing authorisation. The value attached to civil aerospace was reiterated in subsequent plans, and the Eighth Plan of 1980 in particular, contained a commitment to continue the improvement of industrial infrastructure and support for the Airbus and the CFM56 engine. [53]

Outsiders have sometimes attributed more coherence to French civil aerospace than may have been present in the field. The French planning system, and an annual budgetary allocation to civil aerospace, gives the impression of a strong, centralised direction to French civil aerospace. As the British SBAC observed, 'there is no doubt that this consistent strategy to utilise the aerospace industry as one of the mechanisms to success has already achieved results.... The advantages of having a national strategy for aerospace...are clear to the French Government'. [54] However, the planning system only serves to set broad objectives, which can, and have been, undermined by the volatility and uncertainty of the civil market. French companies have also complained that specific programmes have not always received the level of support which they felt was needed. As one industrial reaction to the 6th Plan put it, 'its provisions will severely constrain our ability to develop big programmes'. [55] Similarly, the priorities set by the government sometimes showed poor commercial judgement; at one point in the late 1960s, the Airbus ranked behind the Concorde and the Mercure in terms of political importance and financial provision. In the mid-1970s, organisational changes were made in the French Ministry of Transport, precisely because of problems of policy coordination and a lack of coherence in French civil aerospace decision-making. [56]

Anglo-French Collaboration

Collaboration provides a common policy theme in France and Britain. In the late 1950s, both countries came to realise that the problems of rising costs and limited national markets demanded international solutions. Britain and France have collaborated in a number of civil and military projects. From the late 1950s, some saw them as 'natural' partners, but relations between the two have not been easy. Although there is great respect at an industrial level between the two countries, mutual suspicion and recrimination following the breakdown of some collaborative schemes in the 1960s, left their mark. More insidiously, the British have sometimes felt that generally they have given more to the 'less comprehensive European industries' in collaboration than they have received. [57] The French, for their part, have believed that official and political opinion in Britain has too often favoured cooperation with the US over Europe. To some extent this has diminished with BAe's confirmation as a major partner in Airbus Industrie, but the French can still be irritated by what they regard as Britain's narrow approach to the development of European civil aerospace. British governments, on the other hand, have regarded adherence to commercial criteria as a sensible and necessary precaution against ill-judged, over-ambitious European projects.

In terms of European technology, not just in civil aerospace but in satellite launchers, micro-electronics, and other key sectors, the French do see themselves as the initiators and the catalysts of collaborative programmes. [58] As such, French claims to leadership are not always well received in Britain. This may well reflect a degree of envy, but it does add to the friction sometimes present in Anglo-French relations. The British often appear less whole-hearted in their commitment to Europe and, perhaps, are less willing to accept an unqualified view of a project's commercial viability just to advance European technology. Yet in tandem, the two industries do present a formidable capability, and the future of a European civil aircraft industry, if not a European civil engine industry, largely depends upon the strength and durability of relations between the two industries.

Important as they are, the British and French aircraft industries do not constitute the entire European aircraft industry. Indeed, the West German industry has played an increasingly important role in the development of European programmes and, as so often in European politics generally, Germany has often acted as an arbiter in Franco-British aerospace wrangles. The Dutch company Fokker has an impressive civil aerospace capability, and despite its involvement in Airbus and other European ventures, it has followed an independent path between the larger European companies. In addition, Spain is a full member of Airbus Industrie, and Belgium makes a significant contribution to the Airbus programme. The Italian industry, while it participates in a number of European projects, has a distinct American orientation. Despite several invitations, the Italians have not joined Airbus Industrie, although some Airbus work has been subcontracted by German firms. Aeritalia subcontracts to both Boeing and MDD, and is also involved in the latter's prop-fan programme. [59]

West Germany

Prohibited from building aircraft until 1955, German industry has rediscovered its substantial aviation tradition almost entirely as a result of collaboration, and is now the third largest in Europe. In common with most of its European neighbours, the current structure of the German industry is the result of extensive domestic rationalisation. The airframe industry is now centred on MBB and engine development on MTU. Rationalisation was given additional impetus by collaboration; German firms had to be capable of holding their own with the British and French. The German aircraft industry remains in private hands, but is heavily dependent on Federal and Lände governments for risk capital and for financial guarantees. Up to the mid-1960s, civil programmes were a small part of the overall German aerospace effort. In 1967 only 14% of the workforce was involved in civil projects, and 80% of the industry's contracts were defence related. The Federal Government decided that dependence on public markets should be reduced by expanding the civil sector. In the short term, however, the Government had to increase its support for civil aerospace, and subsidies for civil projects were raised from 60% to 80%. [60]

Germany's share of the Airbus A300 and A310 has accounted for most of this, and has entailed an investment of over $2500 million. In theory, one-third of this was to be provided by industry; but with government guarantees, public support for the Airbus has averaged over 90% since its inception. More broadly-based attempts by the Federal Government to increase private capitalisation of civil projects have been successfully resisted by German companies. The multiplicity of government departments with a responsibility for aerospace, and the interests of the various Lände Governments with shares in regionally based companies made the formulation of a coherent policy even more difficult. [61]

Matters were further complicated by the connection made between German aerospace policy and progress towards an integrated European aircraft industry; 'effort must be made to progress beyond the stage of *ad hoc* cooperation in joint ventures such as the Airbus and to establish an industrial structure which corresponds to the opportunities provided by an enlarged Common Market'. The formation of VFW–Fokker in 1968 seemed to presage a broader realignment of European industry. However, their larger neighbours were not inclined to follow the German–Dutch example, and even Fokker and VFW eventually had to admit that their transnational merger was a failure. German policy, therefore, increasingly came to reflect the reality of nationally-based aircraft industries. The implications for Germany were starkly outlined in the mid-1970s by a quasi-official report into the industry. The civil sector in particular was characterised by 'isolated activities, which have not been incorporated into an overall concept'. German ambitions, the Report added, had outpaced both the market and the emergence of an integrated European aircraft industry. The government itself had to admit that 'initial moves towards a unified European aerospace industrial policy appear difficult to bring to a successful conclusion, because of developments in other countries'. [62]

The appointment in 1974 of Martin Grüner as Coordinator for Aerospace at the Economics Ministry, provided an opportunity for a major review of German policy, as well as bringing much needed centralisation to German aerospace decision-making. His first report, delivered in 1975, outlined a policy framework to last until the end of the decade. He recognised the advances made by German industry since the mid-1950s, and noted that the German contribution to joint ventures was widely appreciated in Europe. Certainly, German withdrawal from current civil projects would have 'far reaching effects on the European and German aviation industries and could only be rectified at a later stage by additional expenditure'. However, if German companies expected to receive more assistance from the German Government, they would have to increase their efficiency and effectiveness. Similarly, collaborative programmes would have to pay more stringent regard to commercial and financial factors. The Grüner Report also anticipated further rationalisation in the German industry, and recognised that this might be incompatible with transnational strategies such as that adopted by VFW–Fokker. In short, Grüner looked for a greater degree of realism in German policy towards aerospace, particularly in civil aerospace, at both national and international levels. [63]

By the mid-1980s, German attitudes towards European cooperation had become rather more pragmatic. Collaboration was still the only realistic approach for German industry, but national aerospace policy was centred on a strong, rationalised domestic industry, implemented through commitments to individual European programmes. As Grüner put it, 'Our aviation policy is based on European cooperation, and the future lies with the programme based companies, Airbus Industrie, Panavia and Euromissile.' [64] West Germany is a solid, if now a less idealistic supporter of European collaboration. It is unquestionably an important actor in European civil aerospace policy, and, as Germany's successful claim for more recognition and authority in Airbus Industrie attests, it is increasingly unwilling to play a subordinate role in joint programmes. [65]

Holland

Like the Germans, the Dutch have been consistent advocates of European cooperation. However, the Dutch have also been conscious of their own needs, and especially to avoid being swamped by their larger neighbours. The Dutch aircraft industry consists almost entirely of one company, Fokker. Although twentieth placed of the world's aircraft companies, Fokker has maintained a strong civil design and development capability. Throughout the 1970s, civil production accounted for between 70% and 80% of Fokker's output. Participation in the F16 fighter programme has helped to maintain cash-flow and to improve the firm's manufacturing technique, but Fokker's main business is still centred on civil aircraft. Fokker's projects have been reasonably successful in commercial terms. [66] However, the limitations of a small domestic market and capital base, led both Fokker and the Dutch Government to seek foreign partners. Fokker was, of course, one half of the Fokker–VFW transnational merger. In the event, the merger proved to be unsuccessful, and the joint company was dissolved in 1980. Even before the collapse of VFW–Fokker, the Dutch, like the Germans, had become increasingly disenchanted with European collaboration, at least in its more ambitious forms. It became an associate member of Airbus Industrie in the early 1970s, but has been unwilling to limit its autonomy by joining the consortium as a full partner. The Dutch also tried unsuccessfully to develop closer links with the US and Japan. In 1983, Fokker launched the F50 and

F100 at an estimated cost of £225 million. This was too high for the Dutch Government readily to absorb alone and Fokker has again been obliged to seek foreign partners. [67]

In short, Fokker has retained a modest but resilient place in the civil market. But it is a small producer in a business where sheer size is an important factor in ensuring corporate survival. Indeed, Fokker's limited capability was a major reason why the Japanese were reluctant to collaborate with the Dutch. Fokker's predicament has been aggravated by its growing ambivalence towards European collaboration. Although reality forces the Dutch to seek partners, there is a real fear of lost independence and autonomy. Fokker has come to prefer carefully-limited cooperation with the smaller European firms, or with the more distant Americans. Fokker's future is, however, dependent upon success in one of the tightest civil aerospace markets. Any major setback may force both Fokker and the Dutch Government to re-think its attitude towards collaboration with the Airbus consortium, or to try again to establish closer links with MDD or Boeing.

The Far East

It is clear that the Far East is rapidly becoming one of the most important markets for civil aerospace. It is equally the case that countries in the region have recognised the technological and industrial significance of developing a manufacturing capability to match their role as a market. Australia, Japan, China and Indonesia are already subcontracting to the established companies of Europe and the US. Many companies are also involved in more ambitious cooperative ventures. [68] Certainly Japan, and perhaps later, China, are set on becoming major forces in the civil aerospace industry.

Japan

In John Newhouse's phrase, the Japanese are the 'wild card' in international civil aerospace. [69] Aerospace, and civil aviation in particular, is one of the few high-technology industries where Japan does not have a substantial presence. In the view of Japanese officials, as newly-industrialising states enter automobile and other medium-technology markets, Japan must move into still more knowledge intensive sectors. As growth in military aerospace is still limited to some extent by political sensitivities towards defence, civil aerospace, along with advanced computers and new fuel technologies, has been singled out as one of the three major elements of Japan's future economic development.

The Japanese aircraft industry consists of five major units, all components of the large industrial conglomerates, Mitsubishi, Kawasaki, Fuji, Ishikawajima and Harima. Aerospace represents only a small fraction of their output (about 15% in total), and they are reluctant to increase substantially the ratio of such a high risk area to their existing interests. In practice, much of the work in aerospace is conducted by consortia based on these five manufacturers. The JAEC consortium is currently involved in the development of the V2500 engine with Rolls and P & W. In 1986, an airframe group, the JCAC, signed a comprehensive agreement with Boeing to develop a 150 seat prop-fan airliner. The consortium approach is dictated as much by financial conservatism as by the need to pool technical and industrial resources. The Japanese Ministry of International Trade and Industry (MITI) has sponsored a cautious but comprehensive civil R & D programme since the mid-1960s. This culminated in the YS11 turbo-prop airliner, which although a technical success, was regarded as an expensive failure. In the light of the YS11 experience, MITI encouraged Japanese companies to form links with larger, more experienced partners, but at the same time, continued to support, through domestic research programmes, the improvement of national skills. [70]

MITI has adopted a long-term view of Japan's aerospace future. In October 1982, it produced a strategy paper outlining the case for a significant civil aerospace capability: 'The aircraft industry is a high value-added industry, technically advanced sector; a typical example of a knowledge-intensive activity which will create a ripple effect throughout the rest of industry. The new aircraft and jet engine projects could play a vital role in the development of Japan's creative, knowledge-intensive industries and the eventual creation of a technology based Japan.' [71] Moreover, MITI has taken a far more direct role in charting the development of aerospace than it has done in other manufacturing sectors. Usually, firms have had to set their own innovation targets and privately finance their own

R & D. MITI encouraged this process through tax incentives and by providing investment and market guidance. MITI's main concern was to build a consensus about future development, and to coordinate industrial programmes, thereby avoiding duplication and unproductive competition. In the aerospace sector, however, MITI has taken a lead in setting national goals, as well as in helping to define specific technological and commerical options. MITI takes a close interest in the progress of individual projects. For example, the JAEC Board has five representatives from industry, one from MITI and a Ministry of Finance auditor. MITI is well aware that a rather conservative industry will need continual and generous government assistance. It usually provides a substantial share of development costs in the form of loans repayable from the profits of individual programmes. Collaboration, preferably with a market leader, is regarded as the best way to acquire expertise with the lowest risk, and as a vehicle for defusing criticism of Japanese protectionism and export penetration of Western countries. The Japanese are well aware of their value as collaborators to established manufacturers and aim to have a role in all phases of design, construction and sale. This combination of ambition and caution has often led to protracted and frustrating negotiations, inevitably complicated by American and European concern to protect their technology.

MITI's long-term aspirations may exceed those of the Japanese aircraft industry. While MITI feels that Japan could become an independent producer, many Japanese industrialists doubt both its feasibility and desirability. They imply that MITI has failed to appreciate fully the risky and complex nature of civil aerospace. The limitations of Japan's domestic market for civil aircraft have deterred industrialists used to building up an export offensive based on a saturated home market, effectively closed to outside penetration. According to Mitsubishi, 'the only way to develop new airliners is with international cooperation' [72] Even within MITI there are some who have doubts about adopting a more ambitious civil aerospace policy. [73] It is, perhaps, still too early to assess the chances of Japan breaking into the civil aerospace business as an independent producer. The odds are that Japan will become a strong partner, able to drive hard bargains with potential collaborators, but without a fully comprehensive aerospace industry. Japan has still to reach the level of West Germany in terms of turnover and capacity, and for many Japanese industrialists this is their primary objective. [74] Much will also depend on whether the constraints on Japan's defence sector are relaxed; if they are, then the Japanese aerospace industry would be able to expand its technology base and increase its profitability. In turn, this would provide a more secure foundation for an independent civil capability. Set against this is the limited scale of Japan's current R & D and production facilities, and the complexities of the international civil aerospace market, with their many barriers to new entrants.

Some Americans believe that independence will probably be beyond Japan's grasp. Tom Bacher of Boeing does not underestimate Japanese skill and energy, but he does not see them 'taking over this industry. I think they are more interested in being a partner than a competitor'. John Newhouse agrees with Boeing, the experience of working in the 'heartbreak market' of civil aerospace might have a salutary effect even on MITI, 'they may recognise that the process of making, selling, and looking after airliners is a far cry from the automobile and consumer electronics business'. [75] *The Economist* is less sanguine, 'the Americans and the Europeans like to think that design skills can be kept secret, despite collaboration. That is what their car companies thought too'. [76] The Japanese appreciate both the fear and the attraction they hold for the likes of Boeing and Airbus Industrie. For the moment, Takashi Yoshinari, from the Society of Japanese Aircraft Companies, claims that Japan only wants an effective industry. In the short term there will be little likelihood of an industrial takeover, and 'effective international cooperation' would help to avoid it in the longer term. [77]

Cynical observers of the civil aerospace industry might readily ask themselves why the Japanese, or anybody else for that matter, would deliberately choose to get involved in the business of building airliners and civil aero-engines when the return on investment often seems to be in inverse proportion to the risk. There is an intrinsic glamour and challenge in developing and selling airliners, 'making and flying a Boeing 727 is inherently more exciting than making and selling paper clips'. [78] Civil aerospace has never been a safe industry, and in recent years, life has got still more difficult. There may be an element of romance attached to building airliners and big civil engines, but this has been steadily tempered by cold economic logic.

Notes

1. D. C. Mowery & N. Rosenberg, 'The Commercial Aircraft Industry', in R. R. Nelson (ed.), *Government and Technical Progress*, New York, 1982, p.102.
2. Ibid. See also, R. Millar & D. Sawyer, *The Technical Development of Modern Aviation*, London, 1968.
3. T. Bracher, *Boeing Commercial Aircraft Corporation*, Feb. 1981, p.10.
4. C. D. Bright, *The Jet Makers*, Kansas, 1978, p.78. See also, R. Doganis, *Flying Off Course*, London, 1985, ch. 1.
5. Millar and Sawyer, op. cit., pp.117–89.
6. K. Hayward, *Government and British Civil Aerospace*, Manchester, 1983, ch. 1.
7. Bright, op. cit., p.100.
8. *AWST*, 22 August 1983, p.41.
9. R. E. Martens, Vice-President Financial Planning and Analysis, *Interavia*, Dec. 1985, p.1344; *AWST*, 3 October 1983, 14 November 1983, p.70.
10. *Flight*, 31 December 1983, pp.722–3.
11. Ibid, 11 February 1984, pp.394–5.
12. A. Phillips, *Technology and Market Structure*, Lexington, 1971, p.126.
13. *AWST*, 13 February 1984, pp.135–40.
14. *Report of the Committee of Inquiry into the Aircraft Industry*, Cmnd 2538, London, 1976, para. 84.
15. See K. Hartley & W. Corcoran, 'The Time-cost Trade-off for Airliners', *Journal of Industrial Economics*, Mar. 1978. The Americans could make the similar mistakes, note for example, Convair's relations with TWA on the C880/990, and Lockheed's problems with the Electra.
16. Sir Austin Pearce, 'The Business of Meeting Aerospace Needs', *Aerospace*, Oct. 1985, p.25; *Flight*, 25 August 1984, p.98.
17. Hartley, op. cit., pp.221–2, See also Chapter 5.
18. *AWST*, 26 September 1983, p.44; *Interavia*, Feb. 1985, pp.127–8.
19. *Flight*, 14 January 1984, pp.82–4.
20. R. Gidwitz, *The Politics of International Air Transport*, Lexington, 1980, pp.207–8. Doganis, op. cit., Ch. 3.
21. Mowery & Rosenberg, op. cit., p.116.
22. In 1981, civil business accounted for three-quarters of Boeing's earnings. By 1984, the proportion of civil to military had changed slightly in favour of defence work, but civil aerospace still accounted for 71%. Boeing is 27th in the Fortune top 500 US firms and third placed aerospace company.
23. *AWST*, 21 November 1983, pp.14–15. Although Douglas was able to match Boeing's venture into jet airliners, the financial strain of developing the DC8 and then the DC9 was a primary reason for its

vulnerability to takeover bids. Interestingly, Lockheed was also interested in merging with Douglas.

24. *Interavia*, 9 April 1984, pp.915–17; *The Economist*, 1 June 1985.

25. Bluestone *et al.*, *Aircraft Industry Dynamics*, Boston 1981, p.39.

26. J. Newhouse, *The Sporty Game*, New York, 1984, pp.111–14.

27. Although GE lost money on the C5A contract, its caution in waiting for the defence contract to cover the initial launch costs of the CF6 gave it a firmer base from which to launch the civil version. The delayed introduction also led to the production of a very reliable engine.

28. See T. Bacher, *The Economics of the Commercial Aircraft Industry*, The Boeing Commercial Airplane Company, Seattle, 1984, p.17.

29. Mowery & Rosenberg, op. cit., pp.140–6. In a recent example, Boeing received a NASA contract to explore the uses of carbon fibre tail surfaces, data which was used in the 757/767 series. GE has also had similar contracts to work on UDF engine research.

30. *Flight*, 12 June 1976, p.1550.

31. See statement by K. G. Harr, President of the AIAA, *Industry Week*, 17 January 1972; see also, *AWST*, 31 May 1971, p.58; Bluestone, op. cit., pp.159–60.

32. *Action Programme for the European Aeronautical Sector*, Brussells 1975, R/2461/75 (The Spinelli Report) p.3.

33. K. Hartley, *NATO Arms Cooperation*, London, 1983, p.141.

34. Ibid., p.142.

35. R. Nobbs, Memorandum to the IPSA/EC Commission Study Group, *The European Alternatives*, 1976, para. 2.6.

36. K. Hayward, *Government and British Civil Aerospace*, op. cit., pp.163–5.

37. F. Franzmeyer, *Approaches to Industry Policy within the European Community and its impact on European Integration*, London 1982, pp.120–24.

38. Hayward, op.cit.

39. *Flight*, 13 May 1978, p.1437, See also Alan Greenwood of BAC, *Flight*, 29 May 1975, p.847.

40. See R. Williams & M. Edmonds, with K. Hayward & M. Dillon, 'Air Policy', in G. Ionescu (ed.), *The European Alternatives*, Leiden, 1980. On the other hand, others want the EC to stay out of the whole area, leaving matters to the free market. See Keith Hartley, *Financial Times*, 28 March 1984.

41. *The Times*, 6 September 1982.

42. Hayward, op.cit., Ch. 1.

43. *Flight*, 6 April 1984, p.3; *Financial Times*, 31 August 1985; *Sunday Times*, 22 September 1985.

44. *Financial Times*, 9 November 1985; *Guardian*, 9 November 1985.

45. House of Lords, *Select Committee on the European Communities*, Session 1975–6, HoL (305) Qs 535 & 560.
46. See Hayward, op.cit., pp.215–22.
47. *Interavia*, August 1984, p.777.
48. SBAC Memorandum to the House of Commons, *Select Committee on Science and Technology*, HC37, Session 1981–2. Sir Raymond Lygo, *Times*, 29 May 1985. See also, Sir Austin Pearce, op.cit., pp.26–7.
49. G. Jalabert, *Les Industries Aéronautiques et Spatiales en France*, Paris, 1974, pp.128–37.
50. C. Layton, *European Advanced Technology*, London, 1969, p.128.
51. Rapport Parlementaire No.2815, cited *Flight*, 14 May 1977, p.1364.
52. Raymond Barre, *Air et Cosmos*, 18 June 1977, p.11.
53. See Preparation de 7e Plan, *Industrie Aérospatiale*, Commissariat du Plan, La Documentation Française, Paris 1976; M. Herman, 'L'industrie aérospatiale française', *Économie et Politique*, Mar. 1966, p.93; Bulletin, *Sociéte Générale*, May 1982; *Interavia*, June 1975, p.614; *Air et Cosmos*, 12 July 1980, pp.12–4.
54. HoC 37, op.cit.
55. Jalabert, op.cit., p.68; *Le Monde*, 8 June 1979.
56. *Air et Cosmos*, 9 February 1980, p.9. One French observer described the administrative relationship prior to the centralisation as 'Kafkian', Phillipe Glazier, *Le Nouvelle Économiste*, 5 January 1976, pp.16–18.
57. *Keynote Business Interests*, Mar. 1983, p.7. For a sense of French attitudes towards collaboration, see statements by Bernard Lathiére, Airbus Industrie, *Interavia*, June 1975, p.616, a French Government spokesman, *Flight*, 12 June 1973, p.44, and J. N. Adenot, President Aerospatiale, *Flight*, 9 November 1972, p.652; Jean Blanchard, French Minister of Armaments, 'Conception et réalisations des armaments', *Revue Défence Nationale*, Feb. 1971.
58. See *Le Monde*, 29 August 1984; and especially, a series of articles by F. Lagrange, Advisor to the French Conseil d'État, *Le Monde*, 15, 16, 17 November 1984.
59. *Financial Times*, 29 July 1985; See *Flight*, 1 June 1985, pp.137–9, for a review of European subcontracting to the United States.
60. *Interavia*, April 1972, p.328.
61. Report from the Federal German Commission for Economic and Social Change, *Interavia*, July 1975, pp.772–3.
62. Ibid.
63. The Federal Minister for Economic Affairs, *Working Party Coordinator for Aerospace*, (The Grüner Report), Nov. 1975, p.15.
64. *Flight*, 13 May 1978, p.1436.
65. Ibid., p.1436.
66. The Fokker F27 is the world's best selling turbo prop airliner, and its

jet successor, the F28, is approaching its break-even point.
67. *Interavia*, February 1984, pp.173–4.
68. China has a co-production agreement with MDD to build MD80s, and will be involved in the development of MDD's propfan programme.
69. Newhouse, op.cit.,p.217.
70. *Financial Times*, 19 September 1983; *Interavia*, October 1983, p.1094; *Flight*, 2 November 1985.
71. *Flight*, 20 November 1982, pp.1521–3; *Fortune*, 1 March 1983, p.109.
72. *Fortune*, 21 March 1983, p.110.
73. *Financial Times*, 19 September 1983; *Interavia*, October 1983, p.1094; *Flight*, 2 November 1985.
74. An idea of Japan's place in the aerospace market can be got from comparing total aerospace sales. 1/30th USA, 1/5th UK, 1/6th France, 1/3rd FGR. *Interavia*, October 1983, p.1084.
75. *Fortune*, 21 March 1984, p.110; Newhouse, op.cit., p.218.
76. *The Economist*, 24 March 1984.
77. *Herald Tribune*, 15 September 1980.
78. *The Economist*, 1 June 1985.

2 Airbus Industrie and the Airbus Programme

Introduction

The Airbus, or now more accurately, the Airbus 'family', represents Europe's main challenge to the American civil aircraft industry, and in terms of total sales, the Airbus is Europe's most successful jet airliner. [1] Airbus Industrie, the international holding company responsible for coordinating development, production and sales, unites the three largest European aircraft companies, Aerospatiale, MBB and British Aerospace. It is also one of Europe's most expensive collaborative projects. Over $7.2 billion has been provided or authorised by the sponsoring governments to cover the cost of developing the three Airbus aircraft, the A300, A310 and A320, of which $1 billion has been repaid. The total cost of the programme, most of which is subject to government guarantee, is estimated to be in the region of $12 billion. The participating governments also cooperate in the provision of export sales credits. Exact details of Airbus financing are clouded in some secrecy, but it is unlikely that the full cost of developing and producing the A300 and A310 will be recovered.

The level of state support for the programme, both direct and indirect, has attracted considerable criticism from US manufacturers. Nevertheless, Airbus Industrie believes that the programme, especially with the A320, will prove to be commercially viable. Beyond any monetary return, supporters of the Airbus contend that it has enhanced Europe's industrial and technological base, and its continued presence in the market removes the threat of an American monopoly in a major commercial sector. In organisational terms, Airbus Industrie's transition from a single project enterprise to one capable of developing a range of products represents a major advance on earlier forms of

European cooperation, and could be the basis of a permanent structure for the development of large civil aircraft in Europe. As will be considered below, Airbus Industrie is not without its problems, and the prospects for future projects are uncertain. However, the emergence of Airbus is one of the most significant commercial and industrial events in the civil aerospace industry since the Second World War. As such, the programme deserves consideration at some length. This chapter begins with an outline history of the Airbus programme, but the bulk of the chapter is devoted to an analysis of the 'Airbus system'. [2]

The A300

The idea of a short/medium range, wide bodied airliner was conceived in Britain and France in the early 1960s. Contemporary political and industrial logic pointed to an Anglo-French project, and between 1965 and 1967 negotiations were held between the two governments and their respective industrial contractors, HSA and Sud Aviation. A German consortium, Deutsche Airbus, joined the discussions in 1966. The airframe companies rapidly established an effective rapport, and by early 1967, they had produced a design outline. The choice of a suitable engine proved more problematic. Rolls Royce, backed by the British government argued that its RB207 design should be the basis for a joint programme. The French, supported by the Germans, preferred a version of the P & W JT9 built under licence. The British Government successfully pressed for the RB207, and in return for this concession, the French were awarded airframe design leadership. In September 1967, the three governments signed a Memorandum of Understanding which committed them to the development of the A300. The agreement stipulated that the three national airlines would each buy 25 A300s, and that no party would support a competing project. Total development costs, including the engine, were estimated at £190 million, with an in-service target set for 1973. [3]

Although the Memorandum had been accompanied by much political rhetoric about the Airbus being the foundation of a European civil aircraft industry, within twelve months, serious political and technical problems emerged. The aircraft's overall weight and size grew substantially, and programme costs rose accordingly. The market, including two of the three 'captive' airlines,

BEA and Lufthansa, was unenthusiastic about the design. Rolls Royce was increasingly preoccupied by the problems of developing the RB211 for Lockheed. More significantly, political support for the project, especially in Britain, was waning. The British Government was also distracted by a domestic alternative, the British Aircraft Corporation's BAC 3-11. [4]

Nevertheless, the Airbus team was given a final chance to produce a viable, and commercially attractive design. In December 1968, they revealed the A300B, a smaller version of the A300, with the implication that it would have a choice of engines. The British Government, by now determined to leave the programme, regarded this as a departure from the terms of the 1967 agreement, and demanded a full review of the project. In March 1969, the French and Germans, tired of British procrastination, decided officially to launch the A300B. [5] A new Memorandum of Understanding, containing strict penalty clauses for unilateral withdrawal and giving the industrial partners greater autonomy to make technical and commercial decisions was signed by the French and German Governments. In December 1970, a *Groupement d'Intérêt Économique*, Airbus Industrie, was formed under French law to administer and coordinate the Airbus programme. The GE CF6 was chosen as the launch engine, with Snecma and MTU as subcontractors. [6] Despite the British Government's official withdrawal, HSA negotiated a private agreement with Airbus Industrie, helped by finance from the German Government. [7]

The political uncertainties associated with its launch did not impress potential customers, and Airbus Industrie had rapidly to establish market credibility. In this respect, the unlimited liability required of its members under the GIE formula and the clear backing of the two governments was especially important. Although the A300 did not have a direct competitor, and market predictions suggested that a high-capacity twin would be needed, the Airbus was three years too early. These difficulties were compounded by the 1973 recession; by March 1974, Airbus had sold only 20 aircraft, almost all to the airlines associated with the manufacturing states. [8] The German Government was increasingly concerned about the rising costs of a programme with apparently little chance of it showing a return. However, the French Government's unwavering support at this juncture for the project and the growing significance of the Airbus to the German aircraft industry, persuaded the German Government to authorise full production. [9]

In 1975, the Airbus at last broke out of its 'home' market with orders from Air India and South African Airways. Indeed, Airbus sales for that year exceeded those of the DC10 and the L1011 combined. Airbus had achieved a substantial lead over any American competitor, a position which was reinforced by the fact that the greater part of its development costs had been incurred before the inflationary surge of the mid-1970s. Undoubtedly, the support from the two governments was vital in protecting the aircraft during the lean years of its initial development. Under a more stringent, privately financed regime, the aircraft might not have weathered the 1973–75 recession. More positively, the Airbus was one of the most technically advanced designs in civil aviation, offering a considerable improvement over existing types. In this respect, HSA's presence in the programme was especially significant, as its wing design proved to be an important reason for the technical success of the Airbus. Without HSA, the programme would have cost considerably more, and the product may not have been as good as it proved to be.

The A310

If Airbus Industrie was to have a permanent place in the market, it needed to emulate Boeing and MDD and produce a family of airliners. By the mid-1970s, attention was concentrated on the 200–20 seat A310. The launch of another aircraft, even a derivative of the A300, would be expensive and the expansion of Airbus Industrie was clearly desirable. [10] An obvious, though not necessarily the easiest prospect, was for Britain to become a full member of Airbus Industrie. The Germans were eager to see the British play a more important role in Airbus, both as a political counterweight to the French, and to improve the programme's financial and commercial base. The 1975 Grüner Report argued that with British membership, Airbus Industrie could be organised 'as a regular European supplier adapted to the world market'. [11] The French were more cautious; although HSA was a loyal and effective partner, Rolls Royce's American orientation was still seen as a potential complication. Nor, for that matter, were the French prepared to compromise Airbus Industrie's place at the centre of the programme.

The British position was indeed affected by the divergent positions of the engine and airframe industries. On the one hand, Rolls continued to regard the American market as its main interest. On the other, the airframe industry, from 1978 largely consisting of the newly-nationalised company, British Aerospace, preferred a European strategy. [12] For its part, the British Government was attracted by the prospect of cooperation between BAe and an American manufacturer, most notably with Boeing. If BAe joined a Boeing programme, not only would it link up with a stronger partner, but with Rolls strongly favoured to supply a launch engine for the Boeing 757, for the first time in a decade, the British civil aerospace industry would point in the same direction. Airbus Industrie was also regarded with some suspicion by British officials. There was a feeling, not wholly unfounded, that the French had structured it to suit their interests. The British also refused to contribute to any of Airbus' sunk costs, even though the new project would be derived from the A300.

The French, supported by the Germans, argued that it was absurd to build a second aircraft outside Airbus Industrie just as the organisation was establishing market credibility. Similarly, with HSA having received £130 million in business from Airbus, the French expected the British to contribute an 'entry fee' of between £100 and £200 million. They were also adamant that British Airways must order either the A300 or the A310. The British Government hinted that it might be prepared to discuss both the role of Airbus Industrie and the principle of some contribution towards A300 development costs, but it could not make any commitments on behalf of BA. Indeed, BA made it quite clear that it did not have a requirement for either Airbus, and would fight any attempt by the government to direct its purchasing policy. [13] During 1977 and the early part of 1978, Boeing offered BAe an extensive array of subcontract work on the 757. As expected, Rolls Royce was to provide the launch engine, and BA hoped to place a big order. Rolls and BA, supported by a strong Whitehall lobby, vigorously promoted BAe's participation in the 757. But BAe was not convinced that its best interests lay in becoming a Boeing subcontractor, and wanted more time to consider European and other American options.

In February 1978, after some uncertainty concerning France's own position in respect of cooperation with American firms, the

French and German Governments decided to launch the A310. Discussions about other projects with other European companies, including BAe, were suspended pending a British decision on the A310. The British Government now came under heavy pressure to decide between Airbus and Boeing. In September 1978, after considerable debate, it was announced that BAe would be allowed to join Airbus Industrie as a full partner. BAe's application was backed by £100 million in loan capital to support its initial share of the A310. £25 million was provided as a payment for Airbus 'work in progress'. Rolls and BA were authorised to make their own arrangements with Boeing, and the necessary financial provisions were made by the British Government. [14] The French, however, were still doubtful of Britain's commitment to the Airbus; as one French spokesman put it, 'I can hardly see how an agreement can be reached if it were confirmed that British Airways is buying the new Boeing 757, which is a rival to the new 200-seat Airbus B10.' An order from BA was regarded as essential proof of British good faith, and a breakdown of talks on this issue was close. A timely order from Freddie Laker was grudgingly accepted as a substitute for BA. [15]

On 27 October it was announced that Britain was to join Airbus Industrie as a full partner from 1 January 1979. Even so, in order further to assure their partners that Britain intended fully to support Airbus, British membership would be subject to a number of temporary qualifications. BAe would become a full member of Airbus Industrie, with a 20% share of the company. As such, the British potentially had a blocking veto in all Airbus decisions, but it was agreed to retain existing arrangements on questions relating to the A300. These restrictions were to last until August 1981, but would immediately lapse in the event of an order from BA. The distribution of A310 profits was also designed to reflect past contributions to the entire programme. [16]

Britain's official return to the Airbus confirmed BAe's position in Europe's most important civil airframe project. Equally, Airbus Industrie was now clearly a basis for most European ambitions in the large civil aircraft production. If Britain had remained outside the consortium, 'it would have had a damaging effect on Airbus Industrie's reputation and re-awakened the memory of previous failures in European cooperation'. [17] BAe's membership of Airbus Industrie effectively removed the chance of a transatlantic 'breakaway' by a major European company. Equally important, the launch of the A310 took Airbus Industrie beyond the single project formula, and represented the first step towards a family of aircraft. [18] By 1980, Airbus had been bought by 40 airlines, and had sold a total of 292 aircraft, with 157 options. Despite this success, overall profitability was still a long way ahead. Airbus Industrie needed over 360 sales to repay direct government sub-vention, and between 900 and 950 A300s and A310s to cover all of its start-up costs. [19]

The A320

Following the launch of the A310, there was a debate within Airbus about the choice of the next project with which to extend further the Airbus 'family'. There was some interest, primarily from the Germans, in a four-engined, long-range aircraft. However, market surveys indicated that a 150 seat, '727' replacement would offer the best prospect. It was predicted that over 3000 airliners in the 140–70 seat category built in the 1960s and 1970s would have to be replaced by the end of the 1990s, a market worth between $50 000 and $70 000 million. MDD and Boeing were also considering new projects to meet this demand, but in the event, both preferred to produce derivatives of existing aircraft as interim contestants pending the emergence of new technologies. Early in 1981, Boeing launched the 737-300, and MDD followed with a variant of the DC9, the MD80. Airbus Industrie decided it had to attack this market with a new, advanced technology A320. The problem was to persuade their governmental sponsors to spend another $1.5 billion while the A300 and A310 were still repaying their launch costs. [20]

In June 1981, the French Government was the first to declare its support for both the A320 and a variant of the CFM56 as a launch engine. [21] The German and the British Governments were non-committal about the A320's 'industrial launch'. At the end of 1981, BAe submitted a detailed plan to the British Government assuming a 20–30% share of the A320 at a cost of £480 million. BAe's request for launch aid was limited to £70–80 million over four years, but the company wanted additional guarantees for loans covering the remainder. The Thatcher Government, determined to reduce the state's role in industry, hinted that BAe would have to find the rest, perhaps from British equipment companies who might stand to gain from a larger British stake in the Airbus programme. The German Government also wanted to reduce the level of public funding in the Airbus programme. Both wished to see substantial launch orders before committing themselves to the A320. [22] Launch customers, however, were scarce; although most airlines agreed that they would eventually need a new 150 seater, money was tight and traffic was recovering too slowly for comfort. The American airline Delta was almost alone in expressing a clear requirement for a new 150 seater, but wanted a new engine such as the IAE V2500 to match an all-new aircraft. [23]

Commercial uncertainty aggravated the political equivocation which surrounded the A320. The French remained dedicated to an early launch. Charles Fitterman, the French Minister of Transport asserted that 'the success of the Airbus programme justifies the determination of the present French Government to support the launching of the A320. I repeat this programme is possible and necessary'. The French Government would do all that was necessary 'to get it built as quickly as possible'. But he had to concede that the decision to go ahead did not lie with the French alone, but required the active support of the other partners. [24] Both the British and the German Governments held fast to the position that the A320 had first to prove its commercial viability.

Early in 1983, there was some sign that French patience might be wearing thin. In a 'leaked' letter to French officials, General Mitterrand, head of Aerospatiale, and the French President's brother, accused the British and Germans of providing inadequate support for the Airbus, and for the A320 in particular. Airbus Industrie moved rapidly to defuse the subsequent row, and the French Government was careful not to press its colleagues on the matter. Sir Austin Pearce, Chairman of BAe, for one, was encouraging but realistic about the A320; 'everybody in the programme is in favour of a family of aircraft', but there were genuine commercial concerns to be sorted out first. Funding for the Airbus could not be allowed to outpace returns on existing products. [25] Bernard Lathière, President of Airbus Industrie, tried to make the best of things. In the interval, Airbus Industrie had incorporated several design improvements, and 'we have mastered the technical, physical, political, industrial, and administrative problems of building an Airbus, but the amount of money involved in deciding to manufacture a third airplane does lead to some decision-making delays'. [26]

The absence of orders, however, had a depressing effect on the project's credibility. The German Government warned of a possible reduction in support for pre-launch development, and the British Government refused to be diverted from a careful examination of all aspects of the proposal. Norman Lamont, the Minister responsible for aerospace, recognised that the A320 was 'vital for the future of the civil aircraft industry in Britain', but 'huge sums of public money are involved.... It is right and proper for us to ensure that the projects are commercial.... We do not want a political aircraft—we want good commercial propositions. [27]

In May 1983, Airbus Industrie was encouraged by the possibility of a £400 million order from BA. However, BA wanted aircraft more or less immediately. Although Airbus Industrie offered to provide second-hand Boeing 737-200s for BA's short-term needs, BA decided to keep its options open by leasing for itself a small number of 737s. BA did not rule out a future purchase of A320s, but this did not help Airbus Industrie's short-term problems. However, in October 1983, it landed a vital customer. British Caledonian ordered seven A320s worth £159 million, with an option on ten more. B-Cal was the first A320 customer outside France, and took total orders and options to 80. More significantly, B-Cal's reputation for hard-nosed business acumen made it the kind of 'quality customer' which the A320 needed. This was underlined by Béteille, 'it proves that a completely independent company, operating on its own money confirms our "product orientated" approach to the A320. The only decision left to take is not that which launches the A320, but that which does not stop it'. [28]

The B-Cal order impressed Whitehall. The DTI, having made its own survey of the market, was reported to be in favour of going ahead. Even more encouragingly, Norman Tebbit, then Secretary of State at the DTI, hinted that the civil aircraft industry might need different treatment compared with other manufacturing sectors. Equally important, in the government's eyes, was the unsettling prospect of a Boeing monopoly in the late 1990s. [29] In November, Sir Austin Pearce briefed Mrs Thatcher and Mr Tebbit, arguing that support for the A320 was not a 'subsidy or free gift'. BAe had already put £12 million of its own money into A320 development, and had every expectation of earning between 4% and 6% on the investment. BAe was willing to find 30% (£300 million) of the money from its own resources, and would repay the government's contribution from the company's general earnings. Sir Raymond Lygo, BAe's managing director, said that BAe would be prepared to take less than its full request if money was 'available up front'. BAe had explored other possible ways of raising the money, but none were satisfactory; according to Pearce, 'we would be broke as a company before we got to the stage of an aircraft being sold'. Without government aid, there would be no alternative but for BAe to withdraw from the A320. [30]

The German Government faced similar problems. Martin Grüner, the Federal Aerospace Policy Coordinator, echoed the British conviction that the A320 had to be a commercial, not a 'political aircraft'. The Ministry of Economics wanted a minimum of between 60 and 80 orders as a basis for support. On the other hand, the government was sensitive to rising unemployment, especially in North Germany, where Airbus work was concentrated. Franz-Josef Strauss, Chairman of Airbus Industrie, a powerful member of the MBB Board, Lufthansa and several German banks, was also pressing hard on Airbus' behalf. In November, the German Government approved a £3 million credit for A320 initial development. Full support, however, remained dependent on more orders. [31] As ever, the French remained the most determined advocates of the A320; President Mitterrand gave his personal support, 'the A320 will be built, and I am its number one salesman'. [32]

By the end of 1983, the A320 had a total of 88 sales and options. Airbus Industrie was confident that marketing and industrial conditions were now ripe for a full launch. [33] Nevertheless, the British and Germans deferred their final decisions until the New Year. The British Government was affected by the spectre of Concorde and the poor post-war record of British civil aerospace. In the words of one executive, BAe had to show that the 'future was not like the past', and that the A320 was not only commercially viable, but an essential element in Europe's claim to be in the technological and industrial 'big league', a status which required prudent, but continual investment. The project was discussed during Mrs Thatcher's Summit talks with President Mitterrand and Chancellor Kohl in January and February 1984, but again she refused to be drawn into a hurried decision. The A320 had to yield 'a financial return commensurate with the degree of risk involved'. The Treasury and some of Mrs Thatcher's more hard line free-market Ministers continued to have grave doubts about the project. According to one observer, Britain's participation in the A320 was 'very marginal' indeed. [34]

The debate in Germany followed similar lines to that in Whitehall. Like the British Treasury, the German Finance Ministry was worried by the risks it entailed, and by the burden on Federal expenditure. Others raised the employment issue and the danger of a Boeing monopoly. Franz-Josef Strauss, although not a minister, was a powerful voice in the governing coalition, and he argued that the A320 was vital 'to maintain our position as a civil aircraft producer'. Chancellor Kohl, if not swayed by the economic arguments, regarded the Airbus as a symbol of Franco-German cooperation, and supported the A320 for diplomatic reasons. On 22 February, the German Government agreed to finance 90% of MBB/Deutsche Airbus's share, worth £388 million. [35]

Only the British Government now remained publicly uncommitted. In fact, the government had already decided in principle to support BAe, but it had to square BAe's needs, the Treasury's dislike for publicly funded civil aerospace, and its own ideological credibility on limiting public expenditure and state intervention in industry. In March, Morgan Grenfell, the merchant bankers appointed to advise on private sector funding for the A320, came up with a package satisfying all of these interests. [36] The government would provide £250 million in launch aid, rather less than the 60% which BAe had wanted, but it was to be given 'front-end loaded', with BAe getting all of its money at the start. The remaining £400 million would come from BAe's own resources and from the City. £50 million of the government's aid would be repaid over three years from 1990, a guaranteed repayment of a fifth of the government's investment. The rest would be repaid through the usual sales levy. There were, however, still some sceptical voices, and one commentator noted that 'every body is happy—at least publicly—except the Treasury and taxpayers, who will not know what has been gained or lost for a decade. In retrospect, the A320 decision may be seen as one which was easier to approve than to refuse'. [37]

Airbus Industrie immediately confirmed the launch of the A320, with formal authority being given on 12 March. The total non-recurring costs of the programme were put at over £2000 million, most of which would come from the sponsoring governments. According to Charles Fitterman, the French Minister of Transport, the outcome was of 'great satisfaction to the French government, and for Airbus Industrie, a truly historic date'. The creation of a family of aircraft was a fundamental stage in the development of Airbus as the world's second ranking airliner manufacturer. It had shown that European cooperation was both 'profitable and durable'. [38] France again had provided the political muscle behind Airbus, and had striven patiently by example and subtle pressure to convince the Germans, but 'chiefly a reluctant Mrs Thatcher', to launch the A320. [39]

The A320 thus became the third major project to be developed under the Airbus Industrie umbrella, a unique record for a European collaborative organisation. The A300 established the programme, and was the first step in a collaborative 'learning curve'. The A310 was the vehicle for getting the British back into the programme, and established Airbus Industrie as the centre for large scale civil aircraft production in Europe. The A320 confirmed that Airbus Industrie was a force to be reckoned with in European aerospace policy. It had played a significant role in the industrial and political lobbying surrounding the A320 launch, helping to maintain pressure on its more reluctant government sponsors. Less directly, the Airbus programme was vital to the health of major national industrial assets in France, Britain and Germany, and as a result stimulated considerable political support. Taken together, the Airbus family would move Europe into a scale of production and investment in civil aerospace hitherto unknown outside the United States. More parochially, French concern about the strength of British commitment lessened with the British Government becoming directly involved in financing the programme. The A320 launch did not necessarily end the political uncertainty associated with new Airbus projects. But the programme had generated a powerful industrial momentum which made it hard, even if clearly not impossible, to resist demands for further investment.

Airbus Industrie

Airbus' impact has been due in no small measure to the success of Airbus Industrie and the Airbus 'system' of development and production. The operation of joint programmes has been one of the main weaknesses of European technological collaboration. Commenting on the experience of the 1960s, Roger Williams noted that 'the particular organisational framework within which (collaboration) comes to be conducted, apart from a monument to the problems being encountered in generating commitment, seems likely itself to be potentially the source of new ones...'. [40] Rene Foch was even more scathing, the management of early joint ventures had 'more in common with the Congress of Vienna than the Harvard Business School'. [41] When the Germans and the French re-launched the A300 in 1969, they were aware of the problems of over-elaborate and intrusive international machinery. The Concorde provided a clear example of how not to organise cooperation in a large-scale civil programme. The two governments were determined to avoid the mistakes of past collaborative projects, and aimed to create an effective and commercially credible industrial organisation with a minimum of governmental interference.

The 1969 Franco-German MoU specified a requirement for an international consortium capable of coordinating all aspects of the programme. However, important national interests, not to mention large sums of public money, were dependent on translating aspiration into reality. As a result, it took over eighteen months of hard bargaining to reach a detailed agreement acceptable to both sides. The debate centred on the extent to which the French would lead the programme. Financially, the partners would be equally responsible, and the Germans wanted to ensure that their input to the programme would be recognised in the organisation. However, in terms of experience and expertise, the Germans were clearly junior to the French, and Aerospatiale naturally wanted to lead on design and commercial exploitation. The privately owned members of Deutsche Airbus also felt at a disadvantage facing the publicly owned and financed French company, especially when the French insisted that financial responsibility for the Airbus should be covered by the total assets of the participating companies. At one point, Deutsche Airbus asked the Federal Government to withhold further funding until the matter was settled. In the event, the

Germans conceded that Aerospatiale would have to have a major, though not dominant role in managing the programme. Deutsche Airbus' financial worries were also eased by Federal guarantees. On 18 December 1970, it was agreed to form Airbus Industrie under French law as a *Groupement d'Intérêt Économique* (GIE). [42]

The GIE formula had been introduced in 1967 to encourage cooperation between French firms. It provided a legal framework for cooperation on individual projects without the participants having to tie up large capital sums in a joint company or to merge. A GIE could be expanded both by the admission of new full or associate members. An associate would assume less of the risk, but would have limited decision-making rights. It was, in short, 'a common tool put at the disposal of the members to render them services needed for the expansion, consolidation and increased profitability of their own activities'. [43] A GIE did, however, demand an unlimited liability of its members, and for Airbus Industrie, this helped to confer industrial and commercial credibility. As one airline executive put it before the formation of Airbus Industrie, 'Boeing has to put its own capital into a product and has to guarantee its customers. But Airbus?' [44] Unlimited liability, backed by inter-governmental agreements against default, provided a firm legal and psychological basis for operation. According to Airbus Industrie, a customer had greater protection in law from a GIE than from a conventional limited liability company.

It would be wrong to view Airbus Industrie as an autonomous design and production centre; primary industrial and technological responsibilities remain firmly in the hands of its industrial members. Airbus Industrie coordinates design, development and production, and each Airbus type has a project manager with a small team to liaise with the partners and with government. But sales and product support are the only functions it performs on behalf of the members. Airbus Industrie is responsible for relations with third parties such as the engine suppliers, and it ratifies the choice of subcontractors made by its members. Associates such as Fokker and Belairbus, are responsible to Airbus Industrie for their warrantees and technical guarantees. But in all other respects, responsibility for the Airbus programme is vested in the 'owners' of Airbus Industrie who also happen to be its main subcontractors. The 'duality' of members-as-owners and members-as-subcontractors could, and sometimes does, cause problems. In practice, much

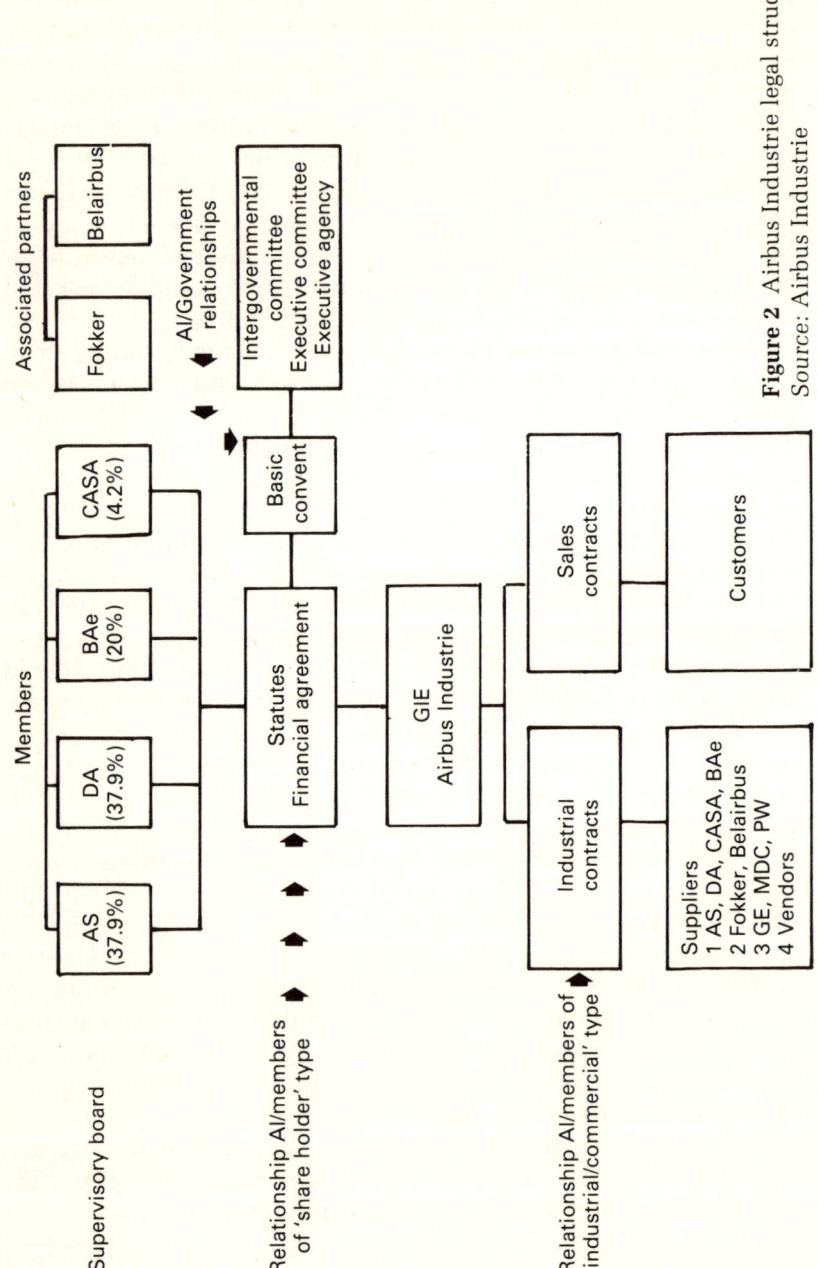

Figure 2 Airbus Industrie legal structure
Source: Airbus Industrie

depends upon good will and the knowledge that the health and effectiveness of a member's business is directly linked to that of Airbus Industrie. To paraphrase Ben Franklin, Airbus Industrie works because if Airbus does not hang together, its members will assuredly hang separately, with Boeing acting as executioner.

The Airbus decision-making structure

The structure of decision-making in Airbus Industrie consists of two parallel systems; one links the industrial partners who are responsible for technical and commercial issues and Airbus Industrie's central functions; the other is a network of official committees which monitor progress and the various Airbus agreements on behalf of the sponsoring governments (See Figure 2)

The industrial structure

Airbus Industrie comprises four full partners, Aerospatiale (37.9%), Deutsche Airbus (37.9%), now almost entirely owned by MBB, BAe (20%), and Casa (4.2%). Shares of Airbus Industrie broadly reflect the financial contribution each partner makes to the consortium as a whole, and determines the voting power of each member. The ultimate decision-making body for Airbus Industrie is the General Assembly of members. The General Assembly receives Airbus Industrie's accounts, ratifies major decisions, such as the launch of new projects, and approves new members. In practice, the functions of the Members' Assembly are assumed by the Supervisory Board made up of 17 industrial members, six from Aerospatiale and Deutsche Airbus, four from BAe, one from Casa and the Chairman of Airbus Industrie, who since its inception, has been Franz-Josef Strauss. It meets twice a year to approve all decisions made by Airbus Industrie. Decisions are by unanimous vote, though Casa cannot veto proposals made by the larger members. Given the routine contact and interaction between Airbus Industrie and the individual firms, the Supervisory Board has rarely overruled Airbus Industrie's management. [45]

As the Board has effectively the same membership as the Member's Assembly, and consists of very senior executives, much of the detailed, day-to-day work of supervising Airbus Industrie is delegated to an Executive Committee. Membership comprises two representatives of the major shareholders, one from CASA and one from Airbus Industrie itself. It meets monthly, and considers issues that affect the programme as a whole. It authorises the 'industrial launch of new projects', it sets the division of work at the start of a project, and adjusts work sharing agreements in the light of developments in the aircraft. The Executive Committee manages the very tricky and often contentious issue of matching production rates to sales. The Board also selects Airbus Industrie's senior management, an issue which, as will be discussed below, has become a matter of some importance. A Financial Committee reports directly to the Supervisory Board, monitoring all Airbus Industrie accounts and budgetary matters.

The duality of the Airbus Industrie system, where its shareholders are at the same time subcontractors to the consortium, inevitably contains a potential for tension and conflict of interest. Both Airbus Industrie and the industrial partners claim that, in general, any source of conflict is matched by the sense of common purpose projected by everybody involved. The success of Airbus is so vital to all concerned, that no company would benefit in the long term from obstructionism and the over-promotion of self-interest. Indeed, company representatives on the Executive Committee have taken decisions that were not to the liking of their colleagues at plant level. [46] Individuals on the Executive Committee, and its superior bodies, take the line that they are acting as shareholders, with a concern for the complete enterprise. They do not see themselves as having 'control' over Airbus Industrie, but rather, as national representatives in an international programme. They do have a responsibility to their respective companies, but they always aim to respond positively to Airbus Industrie's initiatives and to advocate a collective case both within their respective companies and in wider political and public contexts. The key to the Airbus system is 'management by persuasion', decision by consensus rather than by fiat, and consequently formal votes are studiously avoided.

The official structure

There are two levels of governmental involvement in Airbus Industrie. The most senior is the Ministerial Meeting, usually convened twice a year, comprising the national ministers responsible for the Airbus programme. It considers matters such as the admission of new members, and the general status of the inter-governmental agreements relating to the Airbus, and formally authorises the launch of new projects. The role of the Ministerial Meeting tends to be largely symbolic, as major issues, such as the launch of a new project, will already have been a matter for high level political consideration and inter-governmental consultation.

The permanent official system has three elements, the Inter-Governmental Committee, the Airbus Executive Committee, and the Airbus Executive Agency. The Intergovernmental Committee (IGC) consists of senior officials of the sponsoring ministries, and like the Ministerial Meeting, meets biannually. It considers broad questions of policy and related issues. It too formally approves the launch of new projects, but given the political importance of such a decision, this is usually passed on to a higher level. Interestingly enough, discussions about the A320 prior to its launch were not conducted through the IGC as such, but by the same personnel acting in their national capacities. The IGC is more likely to be decisive in matters relating to derivatives of existing aircraft. It also considers general matters (such as the proposed changes in international regulations governing over-water flights by twin-engined airliners) which might be best handled through concerted governmental action. Conversely, it does not scrutinise the Airbus related accounts of the partners, the details of individual subcontracts and prices, nor is it consulted about Airbus Industrie's commercial strategy. The IGC is, however, interested in the overall commercial status of the programme, and is always concerned that Airbus Industrie should behave as a prudent industrial actor. [47] The Airbus Executive Committee (AEC) is in much closer touch with the routine of Airbus operation, meeting on a monthly basis. It oversees the inter-governmental agreements, especially the governmental levies on aircraft sales. It also monitors the work sharing agreements, and examines the effects of aircraft modifications on the overall distribution of work between firms. In short, the IGC, supported by the AEC, is in place to monitor the Airbus programme and to act as a source of information for political decisions about funding and other strategic issues.

The Airbus Executive Agency (AEA), based in Paris, acts as a working level link between Airbus Industrie and the partner governments. Before 1980, it was mainly a French office, but largely as a result of British entry, input from the other states has been increased. [48] Its main functions are to monitor money advanced for development work and the levy on sales. It lacks the authority and direct financial sanction of a national contracting agency such as the British Procurement Executive, and its effectiveness depends upon good will and a close practical relationship with Airbus Industrie. [49] The AEA is also responsible for helping to draw up, and subsequently to oversee the Framework Agreements which form the basis of Inter-Governmental Agreements covering each mark of aircraft. These agreements govern the contribution each state makes to project development. They originate in detailed studies made by the partners and approved by the IGC. The result is an agreed national share for each of the major Airbus types. They provide the basis for the guarantees provided by each government to its partners. Unlike the obligations accepted by the industrial actors, they are not legally binding, but do form a powerful political barrier against unilateral withdrawal. Neither do they constitute national commitments to grant aid to the companies involved in the programme, and each national industrial partner must negotiate the level and form of financing with its own government. The Inter-Governmental Agreements also cover the provision of export sales financing; the governments sign a sales export credit contract for each project, and an official committee is remitted to coordinate export credit support. Different national views about export credit policy have emerged, but unanimity has usually been achieved.

In general, the official network is primarily a communications system, keeping the governments in touch with the Airbus operation, with its most substantive task being to monitor implementation of the complicated Inter-Governmental Agreements. Although the governments are still ultimately responsible for many key Airbus decisions, most Airbus activity and associated decision-making has become routinely a matter for Airbus Industrie and the industrial partners. In many cases, governmental decision is simply a question of vetoing or ratifying choices made by Airbus

Industrie and its members. Compared with earlier cooperative ventures, Airbus Industrie has achieved a high degree of decision-making independence. The growing integration, or at least the common feeling of commercial and technological commitment on the part of Airbus Industrie's members, has generated an increasingly formidable industrial momentum capable of applying considerable pressure on national governments. On the other hand, the protracted and complicated struggle to launch new projects, involving an interplay of national and international politics, remains a significant constraint on Airbus Industrie's ultimate freedom of action and its assumption of full commercial autonomy.

Airbus production

The heart of the Airbus system is, of course, its production network, involving 1400 companies in 17 countries. Within Europe, large sub-assemblies are moved by Super Guppy transport aircraft. Airbus Industrie has pioneered the pre-fabrication of large sub-assemblies, and the relative compactness of Europe ensures that they are in transit for no more than 48 hours. The use of large, integrated sub-assemblies provides the members with a more 'complete' industrial and technical experience, and reduces the mystique of final assembly. As a result, final assembly, although centralised at Toulouse, represents only 4% of the total work. The importance of equalising the technological return to each partner was accepted from the start. As Roger Chanut observed, 'the basic manufacturing requirement was of a sociological nature, that is to recognise the importance for all partners to contribute to the entire range of tasks...'. [50] Airbus Industrie is justly proud of this network, and regards it as being as advanced as any in the USA. Airbus Industrie also claim that the cost of multinational production has been minimised.

Airbus Industrie tries to ensure that production is as efficient as possible, but there is a limit to the pressure that it can bring to bear on the partners. They are not subsidiaries or divisions of an integrated company to be cut or expanded according to demand, but remain discrete national components of an international consortium. The partners have a collective interest in maximising efficiency, but they are also concerned with their own costs and profits. Major changes which might improve efficiency can be expensive. For example, the use of larger metal sections could reduce programme costs, but would entail additional investment by members. At the outset of a new variant, a company might be able to obtain aid, or credit from its government to cover the cost, but this is rarely the case for an established line. Airbus Industrie would have to make out a very good case for such a change, and the savings for the whole programme would have to outweigh substantially the individual costs of improvement. [51] Nevertheless, over the last decade, the companies have considerably improved their efficiency, today taking on targets which a decade ago they would have believed to be beyond their capabilities.

Production and sales matching

One of the more intractable problems for Airbus Industrie is matching output to sales. This is no easy matter for any civil aircraft manufacturer, and working within a multinational framework complicates the process. Airbus production schedules are revised several times a year, and if necessary, voted on by the members. In 1981, Airbus Industrie set a monthly production target of 3.5 rising to 8 by 1984, and perhaps 10 by 1985. [52] The dilemma was, and remains, that when demand rises, Airbus Industrie must maintain production momentum to avoid losing sales through poor delivery schedules, but if sales do not materialise as forecast, production has to be cut or stockpiled. The logical answer is to cut production to match requirements, but the employment practices of European companies, often constrained by political factors, precludes the ruthless 'hire and fire' approach of American firms facing similar cyclical fluctuations in demand. [53]

Historically, inadequate production rates, and consequently poor delivery schedules, have been a fundamental weakness of the European civil aircraft industry. Some Airbus Industrie executives argue that the consortium should stockpile aircraft to back an aggressive marketing strategy, with aircraft ready for dispatch to customers. It is an expensive and risky practice, but was apparently vindicated by the 1984 Panam deal, where immediate availability was one of the factors in the airline's decision. The French have been rather more willing than their partners to support the stockpiling policy, but all have become more sensitive to the costs of over-production. [54] According to one Airbus official, 'the 18 month response time we have developed for Airbus is a major feat as far as European industry is concerned, but we're still miles away from the flexibility of US manufacturers...'. [55] Labour rigidities are still a constraint; for instance, the depression in sales of A300s and A310s during 1984–85, combined with the onset of development activity on the A320 led to a growing imbalance in work load. Aerospatiale, in particular, was affected by the Socialist Government's concern over unemployment, and found it difficult to impose redundancies on production staff while expanding design personnel. Plans for A320 production have taken account of these problems, and Airbus Industrie intends to employ double shifting to cover short term peaks in demand. [56]

Work sharing

One of the fixed points in the Airbus system is the work sharing agreement established at the outset of a new project. The exact national share varies with each type of Airbus, but the broad pattern established by the original A300 has been maintained in the A310 and the A320. When the A310 was launched, and as much of it was common to the A300, there were few problems in adopting a similar distribution. The A320, on the other hand, as a new project requiring a new production line, provided an opportunity to reconsider the pattern of work-sharing. BAe argued that on the A320, and with each subsequent new type, the major areas of responsibility should rotate amongst the partners. In this way, everybody would get experience of every aspect of development and production. BAe wanted responsibility for the nose and fuselage, final assembly and flight testing, hitherto French preserves. Although representing only 6% of the total work and 250 jobs, it would have diluted the 'Frenchness' of the programme. Furthermore, BAe argued that rotation would make better use of its skills and those of its partners.

The French naturally opposed moving the focus of A320 work from Toulouse; 'Toulouse is the natural centre of Airbus activity, customers come to Toulouse, and Toulouse is where the production line will be.' [57] Airbus Industrie objected to the additional cost and inefficiency entailed by duplicating flight testing, sales and support facilities for the A320. Roger Béteille for one was against playing 'musical sub-assemblies'; the key to Airbus' success, he said, lay in specialisation, 'we are strongly against people saying "we have made a wing, now we want to make a fuselage"'. [58] If the Airbus system does contain intrinsic inefficiencies, the development of specialised facilities and talents can only serve to maximise 'learning', bringing savings to the programme as a whole. A more sensitive reason for French opposition was the likely effect on equipment contracting following a major change to the pattern of work-sharing. In the event, BAe's proposal was turned down. The British Government would certainly have had to finance an increase in BAe's overall contribution to the programme, perhaps rising to 35% of the total, a prospect which received little support in Whitehall.

Changes in German requirements further complicated A320 work share negotiations. Initially they only wanted some 15% of the work, a consequence of the German Government's aim of reducing state support for industry. However, as A300 and A310 sales and production failed to increase as predicted, over-capacity in the German industry brought pressure for a larger proportion of A320 work, and at one point, the German claim reached 40%. In the event Airbus Industrie took over a year to settle the broad outline of A320 work sharing, and detailed allocations were still not finalised by the spring of 1986. The launch of the A320 confirmed the original broad division of research, development and production responsibility established at the start of the programme. The A330 and A340 are set to follow the same pattern, and this is only likely to change if there is a substantial alteration in the financial commitment of a major partner. [59]

Responsibility for R & D on the Airbus aircraft broadly follows the main distribution of work. Each partner and major subcontractor is responsible for the R & D associated with its section of work. This is known as 'level two' R & D in Airbus parlance. However, certain areas, such as the flight control system, involve the complete aircraft. This is known as 'level one' R & D, and logically, is the responsibility of one firm. Since 1969, Aerospatiale has had most of the 'level one' R & D. The commercial advantage of having R & D leadership is, to some extent, diluted by liberal technology transfer practices within Airbus. While the details of a new technology or process may be protected, everybody has a general idea of innovations made by the others. However, control over 'level one' R & D is more sensitive because R & D leadership largely determines the initiative in selecting equipment subcontractors. As 'level one' attracts the most important equipment requirements, contract sourcing and the responsibility for choosing suppliers, has become one of the most controversial aspects of the Airbus operation.

Subcontracting and equipment sourcing policy

The equipment issue has two, related dimensions; the distribution of work amongst the member states, and the role of third parties, particularly American firms in the programme. The original Franco-German MoU of 1969 specified that Airbus Industrie had to build an aircraft according to strict commercial criteria, and consequently, to choose components and equipment from the cheapest source, qualified only by user preferences. Airbus Industrie, of course, informs the governments about sourcing decisions, especially when they involve firms from outside the member states and goes to some lengths to account for a contract decision. According to Airbus Industrie, the national origins of a supplier will have a bearing on selection only if all other things are equal. Airbus Industrie takes the view that it should not be seen as a vehicle for the development of European equipment technology. If cost and delivery schedules are to be met, it may have no alternative but to buy from non-European companies.

Specifications and RPFs (Requests for Proposals) are issued to all likely vendors. The company responsible for a particular area of work then evaluates the various proposals and submits its choice to Airbus Industrie for ratification. Airbus Industrie considers the proposal in the light of product support requirements, vendor experience, customer requests and so forth. Once the decision has been made, the designated partner places the order. Airbus Industrie has rarely overruled a member's submission, but if this is the case, the company has fourteen days to accept or reject Airbus Industrie's recommendation. If the proposal is rejected, contract negotiations are re-opened. Clashes between partners are rare, but can occur when 'level one' R & D requirements overlap with those of a 'level two' area. [60] Airbus Industrie policy towards equipment sub contracting is, in short, to create the conditions for a fair and objective treatment of all potential suppliers.

Policy, and the perception of policy are, of course, often two very different animals. Within the member states, the distribution of subcontractors has attracted some criticism, particularly from the British. The fact of the matter is that over the course of A300 and A310 development, French suppliers have provided most of the European equipment for the Airbus. The French were able to use the Concorde as a vehicle for promoting their avionic and aircraft equipment companies, and this policy has, to some extent, continued into the Airbus. [61] Aerospatiale has been responsible for final assembly and systems integration, and was, perhaps, best placed to influence subcontract selection. As a result, French companies were believed to have had an unfair advantage over German or British firms.

When BAe became a full member of Airbus Industrie in 1978, the British hoped to win about 20% of A310 equipment contracts, but only obtained 12%. The poor showing of British equipment companies in the A300 was partly a consequence of Britain's withdrawal from the project in 1969. Commonality between the A300 and the A310 then helped to reinforce the position of existing suppliers. However, French policy in the 1970s was explicitly aimed at building up a French equipment industry as part of a general strategy to improve France's electronics industry. French Governments, in contrast to those in Britain, provided considerable support for the equipment sector, and, as a result, French firms have often enjoyed a price advantage when competing for Airbus contracts. [62] On the other hand, French companies were prepared to take risks and to make the necessary investment in order to win Airbus contracts. Equally, British companies have often been insufficiently aware of the Airbus and the need to come to terms with its procedures. Certainly, there has been a tendency in the British equipment industry to underestimate the skills and expertise achieved by French companies over the last decade.

The launch of the A320, with a defined requirement for radically new equipment concepts, led both the British and the German to demand an increase in their share of equipment subcontracts. The British Government, for example, announced that it would press for a 'full and fair share of contracts', and suggested that a 20%

target would be a more equitable reflection of Britain's contribution to the programme. [63] With the British and Germans set on increasing their share of the subcontract work, clearly something had to give. It was suggested, therefore, that the American content in the A320 would have to be reduced in order to make room for more European products. The A300 and A310 contain a large proportion of US equipment, which including the engine, can take the proportion of American-made components up to 40%. This has provided an important selling point in the United States, and has been used to blunt American political criticism of Airbus market penetration. [64] In the event, the British won nearly 20% of the A320 equipment contracts, and the Germans 34%. Although the exact national origin of equipment for the A320 is being increasingly obscured by joint ventures (indeed, Airbus Industrie has encouraged consortia submission for equipment contracts), the non-European share of equipment contracts in the new aircraft has been reduced to 8%. [65]

The Americans were not slow to attack this loss of business as another politically inspired move to subsidise European suppliers at their expense. For example, under pressure from an industrial lobby as well as from the Department of Commerce, Panam asked Airbus Industrie to increase the level of American supplied components on its order for the A320. Airbus Industrie made some adjustments, and argued that wherever possible, it always tried to match customer requirements, but there was a limit to what could be achieved without incurring substantial financial penalties. Although Airbus Industrie maintains a commercial approach to equipment sourcing, it has taken a strategic decision to reduce its dependence on certain American suppliers both for competitive reasons, and to avoid restrictions under American export control legislation. American suppliers will also have to accept that Europe has made substantial technical and commercial progress in many areas of avionics and systems, and European companies can now match the US for both quality and price. Airbus Industrie's choice of equipment suppliers will inevitably reflect this growing capability. Nevertheless, a reduced role for American companies in the A320 and in future projects will not help Airbus Industrie's case against its critics in the United States.

Airbus Industrie's determination to maintain its policy of buying the best and cheapest equipment from whatever source is a touchstone of its commercial autonomy. It would be naive to suppose that equipment decisions have not been affected by political factors. In general, however, the consortium has achieved a considerable degree of autonomy in this area. Indeed, from a British perspective, it has been able to resist pressure rather too well! If the sponsoring governments want Airbus Industrie to maintain a commercial approach to its operations, they themselves must resist the temptation to influence decisions simply to maintain a principle of '*le juste retour*', or to enforce a policy of European preferences that cannot be justified on commercial grounds.

Airbus Financing

The financial aspects of Airbus Industrie are closely bound up with broader aspects of state support and subsidy to civil aerospace, and as such will be examined in Chapter 5. For the moment, only the internal aspects of Airbus financing will be considered. Airbus Industrie's financial affairs can be divided into three categories; Airbus Industrie's routine payments; the non-recurring costs of programme development; and production funding. The first of these is based on a yearly budget proposed by Airbus Industrie to the members and approved by the Supervisory Board. It includes Airbus Industrie's overheads (about 2–3% of the total) and items of equipment, such as the engines, supplied by third parties. Development and production costs are borne by the members and by Airbus Industrie's associates. [66]

The financial relationship between Airbus Industrie and its members is determined by the GIE formula and its 'fiscal transparency'. Airbus Industrie is not required to publish an annual financial statement, and details of Airbus financing are held by the individual members. Airbus Industrie has been described as a 'mailbox' where money is concerned; it does not hold money, incur 'costs' or, for that matter, declare a formal profit or loss on its own account. In practice, 'the flow of cash and profits or losses between the consortium and its owners is so wholly discretionary as to leave AI's books virtually meaningless if viewed in isolation from the Airbus accounts kept by the owners…'. Airbus Industrie's bankers keep accounts for the four partners, to be drawn on, or deposited in, by Airbus Industrie, in proportion to their overall share of the programme. In total, about 70% of Airbus Industrie's turnover goes through the members' accounts. Airbus Industrie 'buys' parts from the parent companies; it 'pays' for the finished products, which include the costs of bought-in equipment, and reimburses the governments for development costs through payments to the partners. In the future, Airbus Industrie hopes to add net group profits to the partners' accounts. Airbus transactions are made in dollars, and as such, all cash transfers are vulnerable to fluctuations in exchange rates. [67]

The 'prices' charged to Airbus Industrie are based on negotiations between it and its members. These prices provide the basis of an invoicing system linking Airbus Industrie to its contractor-owners. At the outset of each new type or Airbus variant, the members collectively determine the proportion of the total value of the aircraft represented by a piece of work or service. This valuation is derived from detailed studies of the development and production process, and include an assessment of the complexity of the work, commonality with earlier programmes, man-hour requirements, material and other related costs. To take a hypothetical case; if the partners are considering a wing section, each estimates the proportion of the total aircraft cost represented by the wing. A number of estimates result, which, for the sake of argument, might vary between 23% and 28% of the total. The partners might then agree on a value of 25%. The monetary cost of the whole aircraft is estimated which establishes the 'price' of individual sections. Airbus Industrie then 'buys' the wing at this price, subtracting its own overheads. Every ten days, the respective financial obligations of Airbus Industrie and its members are calculated, and invoices, either allocating money to the partners or requesting funds to pay for Airbus' costs, are sent out by Airbus Industrie. At the end of the financial year, Airbus Industrie divides a profit or loss according to the member's share of Airbus Industrie.

Although all of Airbus Industrie's contracts are based on a fixed price, depending on its negotiating skill, or on its dealings with its own subcontractors, a member could make more or less than the defined price on its share of the work. This is not, nor is it required to be, declared as an additional profit. Similarly, the costs and benefits of changes in the exchange rate are the sole concern of the partners. Individual contractors, by increasing efficiency, can also beat their own targets and cost estimates. An official Airbus

Industrie 'loss' on the component could, in theory, be profitable to a member when viewed from its own private accounts. Overt price 'gouging' by a member should be constrained by the common goal of producing an aircraft as efficiently and as effectively as possible, and by the scrutiny of experienced colleagues. The problem for Airbus Industrie is to capture any efficiency gains made by its members. The difficulty being that a company has little incentive to reveal its increased efficiency if this subsequently led to a reduction in the price of its part of the programme. This is, perhaps, an inevitable consequence of consortium development, and a major difference between Airbus Industrie and an integrated, nationally-based company.

In recent years, there have been a number of demands, particularly from the British and German governments, for Airbus Industrie to increase programme efficiency. For example, when the A320 was launched, Norman Lamont, the British Industry Minister, said 'we wish to see not just high numbers of sales, but high numbers of sales at a profit. We would like to see the finances tightened up...'. This has been recognised by the partners. Sir Austin Pearce, Chairman of BAe, observed that it was essential for Airbus to keep a tight hold over costs if it was to compete with a 'single management, single currency Boeing which refers to no one else for its key decisions'. [68] We have seen, however, that measures to increase efficiency are not cost-free or easy to implement, and the responsibility for reducing costs lies largely in the hands of the national contractors. In practice, Airbus Industrie accounts for very little of the direct costs of Airbus development and production. Its only direct commercial responsibility lies in pricing policy. In this respect, Airbus Industrie has been criticised by the governments and by some of its members for setting prices at an uncommercial level. Indeed, the 1985 changes in Airbus Industrie's management structure were in part a result of the members wanting to increase their direct control over Airbus Industrie's commercial strategy. There is, of course, a conflict between prudent financial control and the fight for markets. Prices have to match those of the competition, and during the late 1970s and early 1980s, all three producers were selling on, or below, the margin of commercial viability. But on occasion, Airbus Industrie has appeared more concerned to win a sale than to pay due regard for its cost.

Some financial problems are simply inescapable in an international programme, and the variability of exchange rates is a particularly intractable issue. For example, a ten pfennig change in the Deutschmark–Dollar parity can increase, or reduce costs to Germany by DM500 000 per aircraft. [69] The hardening of sterling against the dollar in the late 1970s and early 1980s, meant that for a period BAe lost money on every set of wings it supplied to Airbus. Cash-flow forecasts are equally affected by variations in the value of the dollar. Between 1970 and 1980, the dollar varied by 30% against an ECU (European Currency Unit) average. Fluctuations on this scale obviously make a difference between a notional commercial 'success' or 'failure', and make it impossible to judge accurately whether transactions are efficient or not.

The GIE structure of Airbus Industrie, and its opaque system of accounting, makes it very difficult for outsiders to formulate objective judgements about Airbus costs. This not only leaves the consortium vulnerable to attacks from its critics, but it can also hinder commercial involvement. For example, in 1984 potential investors in BAe were concerned by their lack of access to the company's Airbus accounts. BAe had to go to considerable trouble to assure them that its commitments to Airbus were manageable and likely to prove profitable. Airbus Industrie and its members claim that there is nothing untoward in the system, and they expect no more confidentiality than MDD or Boeing. But the fact that public as well as private money is involved in Airbus, does raise the question as to whether Airbus must be judged by different and more stringent standards of accountability.

Airbus as a multinational programme

Airbus is quite clearly a multinational programme. Financially, technically and politically, all of the nations involved have contributed greatly to the success of the programme, although the support of the British Government has, at times, been somewhat equivocal. But there is no doubt that the French have maintained a leading position within the programme. The French role in Airbus, combined with their often proprietoral attitude to it, has sometimes caused resentment, even anger amongst its partners. After Britain's withdrawal in 1969, the French assumed *de facto* leadership of the programme, which was reinforced when Airbus final assembly, flight testing and customer liaison were centred on Toulouse. Under these circumstances, a close identification between France and the Airbus has been inevitable. Sir Austin Pearce summed this up nicely: 'customers go to France to see the French assembly and take delivery of an aircraft from a French location'. (70) But if the French have laid vigorous claim to the Airbus, it has been a consequence of their forthright and unremitting support for the aircraft; a faith which has received due praise from its partners. (71)

When Airbus Industrie was established in 1970, there was some grumbling from German sources about the dominance of Frenchmen in senior posts, and that the French would get the most technical and industrial return from the programme. As the Germans grew in experience and confidence, and certainly with the return of the British as full partners, French dominance of Airbus senior management was increasingly at variance with the distribution of work and financial support. The need to widen the national base of Airbus top management had been accepted in principle for some time, and from the launch of the A320, changes were in the offing. For its part, Aerospatiale was sympathetic to the wish of their partners to broaden the national character of senior Airbus Industrie management; as one executive put it, 'when Béteille and Lathière were appointed to their jobs in 1975, the British weren't even a full partner in the Airbus programme, and the whole future of the A300 wasn't clear'. (72)

Late in 1984, Bernard Lathière's application for a renewal of his mandate as President and Chief Executive, and the early retirement on health grounds of Roger Béteille, led to a major upheaval in Airbus Industrie. Lathière, evidently expected to be offered a third five-year term, but under heavy pressure from the Germans, especially Airbus Supervisory Board Chairman, Franz Josef Strauss, his application was turned down. Both Lathière and Béteille had made an enormous contribution to the success of Airbus, establishing it as a credible force in the market and generally lobbying on behalf of the programme, but it was felt that a younger, and more representative multinational team should guide the Airbus as a mature programme. The industrial partners, as well as their respective governments, also wanted to bolster Airbus Industrie's concern for efficiency, with a greater emphasis on pro-

duction management and financial control. After some delay (leading to some speculation about political pressure and conflict
between the partners), Jean Pierson of Aerospatiale was appointed
in March 1985 as Lathière's successor. [73] Johann Schaefler of MBB
became Executive Vice-President/General manager. Although after
only nine months Schaefler left to head Dornier, he was followed
by another German, Heribert Flosdorff. A Briton, Robert Whitfield
was appointed to the new post of vice-president/finance. Later,
another Briton and a Spaniard replaced two other Frenchmen in
senior Airbus posts. [74] Significantly, one of the first tasks of the
new team was to re-structure Airbus Industrie management 'to
increase efficiency, and to enable it to become even more competitive and responsive to customers and the market'. [75]

At a working level, Airbus has long since overcome some of the
worst manifestations of international industrial cooperation. In
particular, the often debilitating struggle amongst the industrial
actors to 'win' project leadership is no longer a significant issue for
the Airbus team. Cynics could be right in asserting that given the
past experience of European cooperative programmes, anything
would have been an improvement. This should not obscure, however, the feeling that Airbus has become a truly European
experience. Felix Kracht, a German veteran of the Transall
programme, and the first head of Airbus Industrie production,
described Airbus as 'Europe at work' with 'French spanners turning English nuts in German bolts'. What counted was ability, 'not
passports'. [76] If the 'high' politics of Airbus can sometimes resem-
ble those of the EC at large, there is less doubt about Airbus
Industrie's ability to do its job—producing a range of effective and
saleable airliners.

The reorganisation of Airbus Industrie is just one symptom of the
fact that the Airbus programme is now clearly more than just a 'one-
off' collaborative project. The need for greater industrial efficiency
over the long term has been accepted by all concerned. There is a
growing consensus, at least amongst some officials, that Airbus
Industrie should begin to assume a more direct supervisory
responsibility for the programme. It is rather anomalous that Airbus Industrie has to set prices and conclude deals with airlines
without having the control over its industrial costs. The new
management team, with more direct knowledge of the industrial
aspects of aircraft development, has already taken a more active

part in internal pricing negotiations between the partner/sub-contractors. By the same token, the partners hope that the new team will adopt a more commercial approach to aircraft sales pricing, a concern firmly backed by their governments. There have also been some discussions about introducing more competition into the pricing of major Airbus components. Although these changes have the widespread support of all involved in the project, they may not be enough for a programme of the size and scale of Airbus, certainly over the longer term.

At an official level there have been a number of informal discussions about the future structure and operation of Airbus Industrie. The French and the British, though the Germans are being considerably more cautious, have even begun to consider the possibility of turning Airbus Industrie into a fully autonomous entity. This would be a revolutionary step, and could imply the power to assign major contracts to companies other than the current partners if they could not match the best commercial price. The political and industrial consequences of turning Airbus Industrie into a 'prime contractor' would be profound, and would effectively turn Airbus Industrie into a genuinely transnational company. Realistically, the chances of such a change, at least over the short term, are somewhat remote, especially given German reluctance even to consider the issue in any formal context. But the fact that such a radical step is even being considered, however hypothetically, is a sign that the Airbus programme can no longer be seen as an *ad hoc* cooperative project, with growth accommodated incrementally.

To date, the Airbus system and the flexibility of the GIE formula has served the partnership well enough. One reason for Airbus' success is the absence of serious competition between the partners in other civil programmes, or, for that matter, in the defence sector. There are some competing projects, such as the Franco-Italian ATR42 and the British ATP, and a larger BAe 146 might edge into some Airbus markets, but no one envisages serious competition in the central core of Airbus business. Nevertheless, the competition which does exist, combined with the strategic problems of effectively merging major national aerospace interests, calls into question any idea of increasing the autonomy of Airbus Industrie to the point of independence. The companies themselves seem broadly satisfied with the structure as it stands; as long as they have to carry the risk of development, they will insist on retaining ultimate control over the consortium. They have accepted the need for improvements in the system, but its basic philosophy seems to be well understood by the participants who know in that in the end survival as major civil producers depends upon recognition of a mutual interest in the success of the Airbus. The question remains whether this will be good enough as the Airbus programme continues to expand.

The decision by Airbus Industrie in January 1986 to launch two new projects, the A330 and A340, has certainly raised important questions about the future structure and operation of the consortium. The A330 and A340 will share a common development programme, as well as drawing on the experience of the earlier Airbus types. This will reduce costs, but full development will still need something in excess of $2 billion in launch capital. Airbus Industrie believe that these aircraft are necessary to maintain and to expand the 'family' concept. Aerospatiale, BAe, MBB, Lufthansa and the German Government want to press ahead, especially with the long range, four-engined A340. But the British Government, and to some extent, even the French Government are reluctant to invest in new projects so soon after the launch of the A320. [77] The British Government, faced by BAe's request for up to £500 million in launch aid to cover the cost of developing a wing and assuming a larger share of overall production, will need a lot of convincing that the claim is justified. Even before Airbus Industrie's decision, the British Government had said it would not provide more money for BAe in the medium term. Although BAe might be able to raise some of the money from private sources, without substantial aid from the government it might have to accept a reduced role in the consortium, and the loss of wing development to the Germans. In any event, Airbus Industrie is also actively seeking new partners to carry some of the risk of A330/A340 development. However, any major shift in the pattern of national contribution to the programme could have a profound effect on the pattern of responsibility and decision-making within the consortium, and could further strain the current organisational structure of Airbus Industrie. Sir Austin Pearce of BAe has suggested that should a new project be launched without a similar pattern of participation, it might have to be managed by a separate company outside the Airbus framework. [78] Shifting Airbus Industrie to an *à la carte* framework, with Airbus

Industrie acting as an umbrella organisation, may require serious consideration. Airbus Industrie has shown itself capable, as in 1978, of absorbing a new partner, and of resolving difficult financial and managerial issues. Clearly, the resolve to find a solution is itself a decisive factor, and all of the partners are broadly agreed on the need to expand the Airbus family. However, a long debate over the A330 and A340, could seriously complicate relations between the partners and threaten Airbus Industrie's hard-won credibility.

The realisation that Airbus is now one of Europe's most extensive, elaborate, and costly collaborative enterprises, has, perhaps, been slow to dawn on those associated with, but not directly concerned in its development. This has been particularly so in Britain where there has not been the same degree of continuous political involvement in the project. Airbus Industrie is now a 'big league' operation, a novel proposition for any European civil manufacturer. Its future depends upon a continuing input of capital and resources to match MDD and, especially Boeing. This, in turn, largely rests upon the willingness of four governments to back the industrial and commercial judgements of their national industries. However, the current round of development choices has undoubtedly increased the pressure on the governmental and industrial partners to think harder about the long-term future of Airbus Industrie and the wider implications of maintaining a European challenge to the United States.

As a final word for the moment on the Airbus, it must always be remembered that in Europe, national political and technological interests are never far from the centre of affairs, and a political dimension inevitably helps to shape Airbus Industrie and the aircraft it builds. Airbus Industrie ultimately stands on a political coalition as well as on an industrial consortium. It was born out of an interweaving of technical and political forces focused on a single project. Airbus has gone beyond this, but it can still be affected by the changing strength of political currents, as much as by the pressures of technological and market forces. However, the Airbus is not going to disappear overnight, and the obligations which bind the participating states are in practice tight and firm. Beyond these, as Lathière himself put it, the Airbus is 'a political symbol of what Europeans can do when they are united—countries that have been trying to kill each other for centuries'. [79] As such, the Airbus has acquired a symbolic as well as an industrial potency which has affected, and will continue to influence commercial and political judgement.

Notes

1. The Vickers Viscount, with 444 sales, remains, as of 1985, Europe's best selling airliner.
2. For a more detailed history of the A300 and A310 programmes, see K. Hayward, *Government and British Civil Aerospace*, Manchester, 1983, and J. Newhouse, *The Sporty Game*, New York, 1982.
3. Hayward, op.cit., pp.77–86.
4. Hayward, op.cit., pp.91–8.
5. In 1970, the new Conservative Government was invited to re-join the Airbus programme. However, the Rolls crisis led both to the rejection of this offer and the cancellation of competing British aircraft, the BAC 3-11. Hayward, op.cit., pp.110–12.
6. See Chapter 4. The Airbus consortium was also able to negotiate a licence with MDD to adapt its DC10 engine nacelle for use on the A300. This saved both time and money, and further increased the Airbus' attractions to DC10 users. With hindsight, MDD might well have regretted being so helpful, but at the time, few in the US thought that the Airbus would be a real challenge to American aircraft.
7. Hayward, op.cit., pp.152–7. Indeed, HSA did rather better than it expected. Profitability was to be achieved after the delivery of 150 wings, but by beating its contractual targets, profits began to flow much earlier. Between 1969 and 1976, HSA earned £60 million on A300 work, and employed 2000 people. The HSA design exploits the phenomenon of super-criticality. This principle seeks to delay the onset of compression and therefore drag at high subsonic speeds. This allows higher cruising speeds and increases flying efficiency and fuel saving.
8. Matters were not helped by the cancellation of an Iberia order. As the Spanish flag carrier, Iberia's order had been considered part payment for entering the consortium. Iberia did later buy the aircraft. Airbus, and the French in particular, were also unhappy that BA bought L1011s instead of A300s. BA claimed that the decision had been entirely commercial, but with Rolls and the L1011 so closely related, this did not appear very believable. Airbus again took some consolation from the diminution of BA's influence over design, and the Airbus could now be more closely tailored to American and the European ATLAS airline group specifications.
9. *Der Spiegel*, 25 January 1971; *Die Zeit*, 17 December 1971.
10. Casa of Spain became a full member of Airbus Industrie in 1971.

Fokker became an associate member in 1970.

11. *Working Party Coordinator for Aerospace*, Report, Bonn, 1975, p.6.
12. Hayward, op.cit., pp.160–71.
13. Ibid., pp.166–78.
14. Ibid., pp.179–80; *Air et Cosmos*, 14 January 1978, p.3.
15. *Air et Cosmos*, 9 September 1978, pp.3, 14; Newhouse, op.cit., p.205; H. Banks, *The Rise and Fall of Freddie Laker*, London, 1982, ch. 11. Laker was allowed to extend his credit even though his gearing was already worse than would normally have been regarded as credit worthy. See also Chapter 3 for detail of US offers to French and British firms.
16. H. H. Schumacher, 'Europe's Airbus', *World Today*, Aug. 1979, p.335.
17. Ibid., p.355.
18. *Fortune*, 21 April 1980, p.141; *Flight*, 27 January 1982, p.1549. Sceptics of the A310 launch have asked why go to the lengths of launching virtually a new aircraft to provide 40 more seats. The A310 is in direct competition with the Boeing 767, and is also subject to pressure at the bottom end of its market slot from the 757. It also costs between $1.5 and $2 million more than the Boeing aircraft. AI's response is that the difference in operating costs more than justifies the A310. See *Interavia*, November 1980, p.1001; *Flight*, 27 February 1982, p.477.
19. *Flight*, 28 March 1981, p.900.
20. *Flight*, 4 February 1978, 27 October 1979, p.1366, 13 September 1980, pp.1069–22, 28 March 1981, pp.900–1. 80% of the 737-300 was common to earlier versions.
21. *Flight*, 20 June 1981; *Air et Cosmos*, 24 May 1980, p.3. See French Minister of Transport, Daniel Hoeffel, *Air et Cosmos*, 7 March 1981, p.11; 'The French Government is ready to give its support to any action which will be necessary for AI to begin the development of a new operation.'
22. *Guardian*, 20 January 1982; *Flight*, 20 April 1982, p.923, 30 January 1982, p.910; *Interavia*, April 1982.
23. *Flight*, 20 February 1982, p.404, 1 January 1983; *AWST*, 6 December 1982, 24 January 1983, p.33.
24. *Flight*, 18 December 1982, p.1734.
25. *Financial Times*, 28 January 1983, 7 February 1983, 22 February 1983; *Guardian*, 28 January 1983; *Flight*, 12 February 1983, p.381; *The Economist*, 5 February 1983 p.69, *AWST*, 30 May 1983, p.178.
26. *Herald Tribune*, 27 May 1983.
27. *AWST*, 6 June 1983, p.30; *House of Commons Debates*, 27 July 1983, col. 1173.
28. For details of the B-Cal deal, see *Sunday Times*, 2 October 1983; *Guardian*, 20 August 1983; *Financial Times*, 3 September 1983, 6

September 1983. See also the remarks made by BA's Chief Executive, Colin Marshall, 'Shareholders in Airbus have to put money behind their faith...Why should we stick our necks out when we have absolutely no need to do so...We have an interim solution leaving our options open...at the end of the decade we will be able to choose what best suits our requirement.', *Flight*, 17 September 1983; *AWST*, 17 October 1983, p.33, 14 November 1983, p.177; *Interavia*, December 1983, p.1334.

29. *The Economist*, 27 August 1983, pp.13–14; *AWST*, 12 December 1983; *Financial Times*, 31 October 1983; *Guardian*, 3 December 1983.

30. *AWST*, 26 December 1983, p.18. One rather optimistic report suggested that either Boeing or MDD might assume Britain's share of the A320. *The Times*, 23 December 1983; *Flight*, 7 January 1984, p.1, 21 January 1984, p.128.

31. *Observer*, 18 December 1983; *Guardian*, 9 November 1983.

32. *Air et Cosmos*, 22 October 1983, p.3, 5 November 1984, p.3; *Flight*, 14 January 1984, p.52.

33. *Financial Times*, 20 December 1983.

34. *AWST*, 12 December 1983; *Financial Times*, 26 January 1984; *Sunday Times*, 22 January 1984.

35. *Der Spiegel*, 27 February 1984, pp.93–6.

36. *Financial Times*, 7 March 1984.

37. *Financial Times*, 2 March 1984, 7 March 1984.

38. *Le Nouvelle Economiste*, 5 March 1984, pp.66 & 69; *Air et Cosmos*, 17 March 1984, p.3.

39. *Flight*, 24 March 1984, p.738; *Air et Cosmos*, 17 March 1984, p.9.

40. Roger Williams, op.cit., p.94.

41. R. Foch, *Europe and Technology*, Atlantic Papers, No.2, 1970, p.52.

42. M. A. Lorell, *Multinational Development of Large Aircraft*, RAND R-2596, 1980, p.65.

43. Airbus Contracts Division, *The Legal Organisation of Airbus Industrie*, Toulouse, May 1980, p.5.

44. *Die Zeit*, 1 June 1979.

45. See Bill Sweetman, *Flight*, 28 March 1981, p.900.

46. For example, when a strike at BAe delayed deliveries of wing units to MBB, the German firm could not fulfil its responsibilities. The Executive Committee decided that the other partners would incur overtime payments in order to deliver aircraft on time. Compensation for MBB was worked out in bilateral negotiations with BAe.

47. A typical agenda for a meeting of the IGC consisted of; progress of forecast timetable; achievement of technical objectives; contracts between Airbus Industrie and engine companies; status of equipment selection; commercial status; sales financing; and progress in drafting

the Inter-Governmental Agreement for the A320.
48. The Spanish have not yet decided whether it is worth having a full-time presence in the AEA.
49. A typical case for the AEA was the status of the two or three A300s set aside for development tests. At the end of the testing, Airbus Industrie refurbished them for sale. AEA had to sort out with Airbus Industrie the proportion of the cost of building the test aircraft attributable to development funding and that under the heading of production finance. A levy on sale then had to be made, with allowances for inflation and exchange rate fluctuations over the period of testing.
50. Roger Chanut, Airbus Industrie Vice President for International Affairs, *European Management Journal*, Vol.3, No.2, 1985, p.71.
51. Even Boeing, a company which invests extensively in systems improvement, will hesitate before changing an established line and mode of production.
52. *Le Monde*, 10 November 1979; *Air et Cosmos*, 17 November 1979, p.11. Airbus Industrie also has to work with labour costs on average 20% higher than in the US. Airbus argue that its productivity is helped by a more extensive use of NCM's. However, comparative costs is a controversial subject within Europe and when Airbus Industrie is compared with US companies.
53. Some at Airbus Industrie believe that labour stability is an advantage, enabling the development of a higher skilled labour force.
54. *Financial Times*, 21 October 1983. The cost of 24 'white tails' during 1984 was estimated at Fr200 million.
55. *AWST*, 31 October 1983, p.30.
56. *Interavia*, May 1985, p.479; *AWST*, 29 July 1985, p.29.
57. *Le Nouvelle Economiste*, 8 June 1981, p.50.
58. BAe has now invested £4.5 million in A320 wing plant and machinery. *AWST*, 6 September 1982, pp.87–9; *Flight*, 6 July 1985, p.44. The Germans are putting up a strong bid for the A330/340 wing in the event of BAe failing to raise the necessary capital.
59. The work share for the A320 is Aerospatiale 36%, Deutsche Airbus 32%, BAe 26%, Casa 6%.
60. Aerospatiale and MBB had a long fight over who was responsible for letting contracts for the control surfaces.
61. See *Air et Cosmos*, 17 March 1984, p.9. On the Caravelle, the French supplied 15% of its equipment (excluding the engine). On the A300, they supply about 50%.
62. Dunlop withdrew from the competition to supply brakes for the A310/320 citing 'predatory financing' in Messier's bid for the same business.
63. *Sunday Times*, 18 June 1984. A claim echoed by Sir Austin Pearce, who called for a more equitable distribution of contracts, to make the

Airbus 'more European'.

64. *Financial Times*, 2 July 1985; *AWST*, 19 March 1984, pp.29, 31; 3 September 1984, pp.151, 153. Airbus Industrie also felt that by reducing the American input, there would be fewer difficulties in selling Airbus to countries in the face of American political objections. The US Commerce Dept noted that even with a Rolls engine, there would be enough US made equipment in Airbus to come under US law.

65. For example, Thompson and Lucas jointly took over the Air Equipment division of Bendix (France) to make equipment for the Airbus; Sfena of France leads a group with a German company and Sperry of the US to develop the A320 flight management system. *AWST*, 3 February 1986, pp.76–81; *Flight*, 18 January 1986, p.50; *Observer*, 9 June 1985; *Financial Times*, 2 July 1985. The national distribution of equipment contracts, with 90% allocated was:

	Fr	Br	G	Others
A320	39%	19%	34%	8%
A310	47%	11%	30%	12%

66. *Financial Times*, 24 April 1985; Airbus Industrie Contracts Division, May 1980, pp.13–14.

67. Ibid.

68. *AWST*, 12 May 1979, p.16; *Flight*, 24 March 1984.

69. *Flight*, 21 April 1984, p.1115; *Interavia*, April 1980, p.309; *Süd Deutsche Zeitung*, 27 December 1978.

70. *Flight*, 21 April 1984, p.115.

71. *Flight*, 27 June 1982, p.1996.

72. *AWST*, 25 February 1985, p.31.

73. See *Le Nouvelle Economiste*, 8 February 1985, p.53. Here the British were described as 'les trouble-fete'.

74. *Financial Times*, 30 January 1985; *Sunday Times*, 10 February 1985; *Die Zeit*, 8 March 1985; *Le Nouvelle Economiste*, 1 February 1985, p.58; *AWST*, 11 March 1985, p.32; *The Economist*, 16 May 1985, p.75; *Flight*, 9 February 1985, p.5; 16 February 1985, p.2.

75. *Flight*, 25 May 1985, p.5; *Financial Times*, 16 December 1985.

76. *Die Zeit*, 1 June 1979. Airbus Industrie exploits its multinational character when selling aircraft to difficult markets. Airbus Industrie invariably sends a Frenchman to Dublin, and the British tend to look after Australasia, not only because of the Commonwealth connection, but also because of French nuclear testing in the region.

77. *Flight*, 1 September 1984, pp.390–1; 16 March 1985, p.3; *Le Monde*, 7 September 1984; *AWST*, 4 March 1985, p.26, 10 February 1986,

pp.48–9, 17 February 1986, pp.27–8, *Financial Times*, 3 June 1985; *Le Monde*, 9 November 1985; *Financial Times*, 9 October 1985, 27 November 1985, 6 February 1986.

78. *Flight*, 18 January 1986, p.1, 8 February 1986, p.6; *Financial Times*, 26 February 1986; *AWST*, 10 February 1986, pp.48–50.
79. *Herald Tribune*, 27 May 1983.

3 Boeing, MDD and Collaboration

During the 1970s, American companies energetically sought collaborative agreements with many of the major European firms. If any one had been successful, it could have had a profound effect on the structure of both the European and the international civil aerospace industry. Although neither MDD nor Boeing was successful in 'picking off' one of the Airbus states, many links were formed between the smaller European aircraft companies and American manufacturers. Certainly the Dutch company Fokker has pursued an independent strategy, associated with, but not fully part of Airbus Industrie. Finally, during this period, the Japanese were increasingly seen as important collaborative partners.

The 'American offensive'

The 1973 energy crisis and its recessionary aftermath, seriously disturbed and delayed the normal cycle of airline procurement. For nearly three years, the market for civil aircraft was severely depressed. This, and the onset of high inflation in most of the industrialised states, had a dramatic effect on the willingness of aircraft manufacturers to accept the risk of launching new projects. Companies on both sides of the Atlantic were unsure of the correct strategy to follow. The market for new civil aircraft would eventually recover, but the problem facing manufacturers at the time was to choose between extending the life of existing designs, and launching new designs. Developing derivative aircraft would be cheaper, but these would be vulnerable to new-technology airliners offering substantial improvements in direct operating costs. Launching a new project, however, carried the grave risk of committing large, and uncertain sums of money to an aircraft which might not enter the market at precisely the right time with the right mix of design features.

Inevitably, most firms tried to delay decisions on new ventures until market requirements were better defined. Design studies proliferated on both sides of the Atlantic. The Americans had been caught napping by the growing attractiveness of the 'big twin', then represented only by the A300 which, as market conditions improved, began to accumulate a substantial following. The American 'Big Three', Boeing, MDD and Lockheed, promoted design concepts in a bewildering array of configurations and designations. The Europeans, either within the Airbus organisation, or in design consortia on the fringes of Airbus Industrie, were equally fertile. [1] Boeing was especially active, touring airlines and prospective partners in Europe and the Far East, with a whole series of ideas. Boeing had been considering 707 and 727 replacements for some time, but the uncertainties prevailing in the early and mid-1970s, forced Boeing to hedge its bets. Indeed, Boeing's generic '7X7' family were described as 'rubber aircraft', with their specifications changing 'according to the airline the company (was) talking to'. [2]

By the mid-1970s, Boeing had more or less settled on two related projects, the 180 seat 7N7, and the larger 7X7 (these would eventually be launched as the 757 and 767). Neither, however, was fully defined until 1978. The company was interested in attracting risk-sharing partners for one, or both of its new projects. Even Boeing recognised that the cost of launching two new aircraft alone would be prohibitive. For MDD, the need to find a partner was more than just desirable; at one point in the early 1970s, collaboration seemed to offer the only way for it to stay in the civil business. [3] Lockheed also considered various developments based on the L1011, and promoted a number of cooperative ventures. However, during the 1970s, Lockheed's interest in new civil projects diminished, and eventually it abandoned the civil market entirely to concentrate on the more certain and profitable defence sector.

From the early 1970s then, American airframe manufacturers began actively to seek collaborative partners. As John Newhouse put it, 'Boeing and McDonnell Douglas each saw a need to acquire a foreign policy.' [4] Their motives comprised a mixture of commercial and political interests. In the first instance, purchases of airliners were increasingly linked to industrial offsets. Secondly, American sources of launch capital were drying up and American firms were attracted by the prospect of obtaining access to European money, especially if backed by government guarantees. Thirdly, American manufacturers were conscious that the consolidation of a European trading bloc might obstruct their free entry to the European market unless European companies were involved in American projects. While these were positive reasons for collaboration, there was also a chance that they might be able to undermine the Airbus by picking off European manufacturers who, for a variety of reasons, were becoming disenchanted with European collaboration. In this respect there was enough uncertainty about the Airbus programme and other projects, as well as domestic pressures in both the French and British industries, to provide opportunities for American 'Trojan Horses'.

Boeing made the first substantive move to attract European risk-sharing partners. In 1971, Boeing and the Italian company Aeritalia, reached a preliminary agreement to build a quiet, short haul aircraft (QSHA). At the time, Aeritalia was under strong pressure to join the Airbus, but the offer from Boeing, which also included the prospect of working on the 707/727 replacement, proved more attractive. The Italians sent a large team of engineers to Seattle to work on the project, which was largely financed by the Italian Government. The inflow of development capital came at a convenient time for Boeing as the company was in serious financial trouble following the launch of the 747. The QSHA programme enabled Boeing to keep valuable design teams together and prepared the ground for future developments.

As Boeing recovered from its crisis, the QSHA became less important. The agreement with Aeritalia was far from solid, and Boeing abruptly terminated the programme. This caused much resentment in the Italian aircraft industry, and left them without a civil project. Aeritalia was offered work on the Boeing 7X7 project, but the Italians would no longer be an equal partner, but 'a small frog in Boeing's great pond'. [5] However, the Italians decided that participation in a Boeing aircraft, even as a subcontractor, was commercially and industrially more promising than a share of the Airbus programme. Despite this experience, the Italians went on to expand their links with both Boeing and MDD. The Italians have also stayed out of Airbus Industrie, despite periodic pressure from the French Government. An invitation to take part in the A320 programme was turned down on financial grounds and because of reservations about the project's commercial prospects. Because of their connections with the American aircraft industry, the Italians have been less enthusiastic generally about European cooperation in civil aircraft, and have opposed the more anti-American implications of EC policy initiatives.

Aeritalia's experience of negotiating with Boeing, and its rapid demotion from a full and equal partner to subcontractor status, did not go unnoticed by other European companies. Boeing might have obtained a foothold in Europe at the expense of the Airbus consortium, but more important companies than Aeritalia noted Boeing's ruthless and single-minded approach to partnership. [6] Nevertheless, and the Italian example notwithstanding, under the right conditions, the major European aerospace nations could still be tempted by cooperation with the Americans. However, interest in transatlantic partnerships tended to reflect frustration with progress in European cooperation rather than any positive commitment to collaboration with American companies.

The Americans and the French

For all their professed determination to take on the Americans, usually couched in terms of encouraging European industrial and technological independence, the French have always looked first to their national interest, and the health of their own aircraft industry. For various reasons associated with failings in European cooperation, from the late 1960s, the French accepted that partnership with an American company would be the best way of building up their civil aero-engine industry. This might also have applied to the French airframe industry. In the mid-1970s, both Aerospatiale and Dassault faced severe financial difficulties, and were working considerably under capacity. Dassault especially was carrying the burden of the commercially-ailing Mercure project. At the same time, the launch of successor to the A300 was encountering difficulties and the whole programme was in danger of stalling. Both Aerospatiale and Dassault were discussing future projects with other European companies, including the British, and some progress was being made towards defining a range of 130–70 seat designs. [7] The main problem was that logically, these ideas would have to be consistent with the Airbus programme. The British in particular, appeared reluctant to join Airbus Industrie, and their position was further complicated by the protracted struggle to nationalise BAC and HSA. Although British industrialists expressed their interest in a European project, there was little sign that specific commitments would be forthcoming at least until BAe was formally in being. The French Government also wanted to promote the Snecma–GE consortium, and was looking for an aircraft project which could be associated with the CFM56. As a result of this evident inertia, the French Government was increasingly receptive to overtures from Boeing and MDD.

In June 1975, the French Prime Minister, Jacques Chirac sounded a warning to France's European partners. France, he said would continue to work for a European aircraft industry, 'but it seems that France's aircraft industry alone in Europe is showing a real will to develop a European capability. The French Government will support the dynamic actions of its aircraft industry even more strongly than before. If necessary, she will turn to other partners in other places'. A French official added, 'French policy is to maintain a living industry in Europe, but not at any cost'. [8] To add substance to French statements, Aerospatiale opened preliminary negotiations with Boeing over a number of alternative ideas centred on the Boeing–Aeritalia 7X7. Dassault soon followed suit, opening discussions with MDD about projects based on a larger version of the Mercure.

In April 1976, Aerospatiale announced an agreement in principle to cooperate with Boeing on a number of related projects. This was outlined in three letters of intent; the first related to Aerospatiale involvement in Boeing's 180 seat 7N7; the second linked the 7X7 with the Airbus B10 (A310); the third implied the creation of a single consolidated programme. Aerospatiale blamed British and German procrastination over the Airbus for its interest in cooperating with Boeing. As General Mitterrand, Aerospatiale's President put it, overtures to European companies 'have not to my knowledge, brought any really positive responses'. [9] Although Lord Beswick, then Chairman designate of BAe, made every effort to convince them that BAe wanted to extend cooperation, the French clearly wanted firm decisions, not just worthy sentiments.

In practice, the letters of intent signed by Aerospatiale and Boeing were far from firm contracts; there was no formal commitment on either side, and both could withdraw at any time. In February 1976, Aerospatiale signed two of the three letters of intent, but until all three were confirmed, there would be no formal agreement. Aerospatiale was happiest with the prospect of cooperating on the 7N7, which promised a large share of the production on a potentially high selling aircraft. On the other hand, Aerospatiale, and the French Government for that matter, were not happy with Boeing's insistence on a commitment to merge the 7X7 and B10 once the 7N7 was in production. In return for a share of the 7N7, Boeing clearly wanted to have some influence over the Airbus. In this respect, French proprietorial attitudes to the Airbus led to difficulties with their partners in the Airbus programme. The Germans were distinctly unhappy about Aerospatiale's contacts with Boeing. They had been given no forewarning of the letters of intent; nor were they consulted by the French about the implications for the future of Airbus Industrie that would follow an agreement with Boeing. The Germans, in fact, believed that Boeing was simply playing games with the Europeans in order to sow dissent amongst the Airbus partners and, more importantly, to pre-empt further European expansion of Airbus Industrie. [10] Although Aerospatiale and Boeing continued to discuss cooperation, any deal involving Airbus began to look less likely.

Even as the prospects for cooperation between Aerospatiale and Boeing began to fade, the MDD–Dassault partnership began to look more substantial. The French Government still seemed determined to look outside Europe in order to promote civil aircraft development. MDD's proposals, which covered a range of civil and military collaborative projects, were particularly attractive. The core of the package was a joint programme with Dassault to build a larger version of the Mercure, combining features of MDD's own 150 seat, ASTR concept. The Mercure '200' would also use the CFM56 engine. In June, following a series of Ministerial meetings, President d'Estaing intervened to press for a transatlantic solution to France's industrial problems. A series of 'basic disillusionments with France's European aerospace partners' were cited as the main reasons for French actions. In particular, the French were disenchanted by Britain's preoccupation with domestic industrial issues, and its government's negative attitude towards Airbus Industrie. As a result, Dassault was virtually instructed to reach an agreement with MDD. [11] In August, Dassault duly announced that it would lead a joint programme based on the Mercure 200, costing $133 million. Although Dassault was officially MDD's partner,

Aerospatiale would have the bulk (40%) of French production. As a nod towards 'Europeanism', Dassault suggested that other European firms would be able to join the project at a later stage. The French hoped to create a risk-sharing consortium with MDD, perhaps linking development of the Mercure 200 with Airbus Industrie and the Airbus family.

Dassault's European colleagues, especially those in the European 'Group of Seven' design consortium were both sceptical and angry about French actions. An original 'Group of Six' European manufacturers had expanded its membership to involve Dassault, but the French company had gone ahead with the MDD deal without consulting any of its erstwhile partners. Although the British had some sympathy for France's industrial problems, they did not underestimate the potential danger to European, as well as to their own interests, created by the Dassault–MDD proposal. [12] The deal had also come as an unpleasant surprise to Boeing, which was still hoping for a substantive agreement with Aerospatiale. Of the two options, however, MDD's proposal appeared to be the better. In the first place, Boeing was not prepared to offer Aerospatiale a genuine and equal partnership. The outline agreement only provided for a risk-sharing subcontract relationship, and Boeing intended to retain total control over key technical and commercial decisions. Secondly, the Boeing offer was unlikely to include a satisfactory role for Airbus Industrie or for the Germans. In short, Boeing and the French were 'a bad fit'. [13]

MDD, on the other hand, seemed more willing to consider an equal partnership with Dassault. Sanford McDonnell claimed that the Mercure 200 would be built 'on the basis of cooperation and not along the traditional lines of a sub-contract relationship'. The French Minister of Transport, Marcel Cavaille was clearly impressed, and said that the link with MDD was 'a choice for national independence, since Dassault will be the programme leader and the engines will be CFM56s developed by Snecma and General Electric'. [14] Aerospatiale, however, was less enthusiastic about joining the Dassault–MDD Mercure 200 programme. Aerospatiale had less confidence in MDD's commitment to an equal partnership, and believed MDD would also want to have full commercial control over any new aircraft. In particular, MDD would want to be in a position to veto any competition with the DC-9 family. [15] With just 15% of the Mercure 200, Aerospatiale suspected that MDD intended to concentrate on its larger DC-X200 project, a potential competitor to the Airbus. Equally, although Aerospatiale would have most of the production work, participation in the Mercure 200 would not enhance its design or commercial skills. According to Yvres Barbré, Deputy Director of Aerospatiale, it was essential for his company to have 'a presence in world marketing'. While he was diplomatic about cooperation with Dassault, Barbré had grave doubts about the Mercure 200's commercial prospects. More fundamentally, Aerospatiale regarded itself as the most experienced civil manufacturer in France, and was somewhat dismissive of Dassault's abilities in the civil field. With the collapse of its deal with Boeing, the Airbus was felt to be Aerospatiale's 'foundation for the future', and Aerospatiale was now determined to extend the Airbus family, with, if possible, BAe as a full partner. [16]

For all the expression of public confidence from Dassault and the French Government, MDD remained non-committal about confirming the preliminary agreement. It was 'gratified' by the French proposal, but far more detailed work was needed before the project could be officially launched. MDD was also very keen to include other European companies prior to starting the programme. Significantly, even French officials conceded that there was no binding agreement between MDD and Dassault, and that other options were still being considered. Moreover, and despite President d'Estaing's public statement, no 'formal policy decision' had been taken. [17] By November 1976, the French Government appeared to be having second thoughts. A firm decision on the Mercure 200 was postponed amidst reports of strong pressure from the British and German governments. A French Parliamentary committee also concluded that the French Government would lose $1.2 million on every aircraft even if 300 were sold. Officials hinted at growing doubts about MDD's offer; 'we are not sure that US cooperation is the best way. It happened last year that it appeared to be the best way. The US way is full of risk and very dangerous. A lot of other things are still possible'. [18]

There were two fundamental problems with the MDD venture. On closer inspection, the potential market for the Mercure 200 looked increasingly dubious. Secondly, the French could not get assurances from MDD that it would not produce a direct competitor to the Airbus. These doubts were enough to shake the French

Government's confidence in achieving a satisfactory deal with the Americans. The French were again constrained by their commitments to other programmes, and the pressure of European politics. The Germans were clearly unwilling to follow the French into a transatlantic programme without firm guarantees. Closer to home, neither of the American options could satisfy broader French industrial and technological ambitions. Both MDD and Boeing had showed a very obvious determination to control all aspects of development, production and sale. If MDD had been less abrasive, it still shared Boeing's single-mindedness on this issue. In May 1977, the French Government broke off negotiations with MDD, and instructed Aerospatiale and Dassault to find European partners. Raymond Barre, the French Prime Minister, now argued that 'the way to equal cooperation with American companies from past experience follows from the strengthening of European cooperation'. The French Government would, henceforward, and with 'renewed determination', support the growth and extension of the Airbus programme. [19]

The collapse of France's transatlantic aspirations has various explanations. Some British observers suggested that the whole exercise contained a strong element of bluff; a ploy to force France's European partners, particularly the British into making a decision. There may have been some truth in this view, but it does seem rather far fetched to see all of this as an elaborate plot to get Europe moving in a direction preferred by the French Government. But the domestic industrial pressures underpinning French actions were considerable, and if they had made a substantive breakthrough with MDD, and launched the Mercure 200, it would have had unpredictable consequences for European civil aerospace cooperation. On the other hand, French internal industrial politics helped to undermine the MDD–Dassault axis. Dassault, although a formidable defence contractor, had limited civil experience, and it certainly needed Aerospatiale to produce the aircraft. For its part, Aerospatiale had little interest in the Mercure 200, and was, in any case, loath to accept Dassault's leadership in a civil project. [20]

The most important reason for the failure of both proposals was that neither of the two American companies was prepared to welcome a concept of collaboration acceptable to the French. MDD and Boeing would have liked access to the capital that seemed to be available from the French Government, but were unwilling to compromise over industrial and commercial control in order to secure a firm agreement. From the French side, although European cooperation could be frustrating, the Airbus, with or without the British, was a more certain option for French industry. The alternatives offered by MDD and Boeing were always likely to be a gamble for France. Neither of the American firms guaranteed participation in future programmes. If the Airbus programme had been seriously compromised, and if, at a later date, something had gone wrong with a Franco-American project, the French civil aircraft industry would have faced serious difficulties. The Airbus, on the other hand, offered a surer route to a 'family' of designs with far more French control over development. The Airbus was also more likely to generate wider benefits for the French equipment and electronics industries than an American programme. With hindsight, French officials admit that the Mercure 200, as well as the Boeing proposals, were a diversion, and that the consolidation of Airbus Industrie was always the most logical direction for French civil aerospace to take. However, at the time, the threat of a French defection was real enough, and was taken very seriously by France's European partners.

The Americans and the British

Throughout their discussions with the French, the Americans were always interested in other cooperative proposals, and during their talks with Aerospatiale and Dassault, MDD and Boeing were sounding out British and other European companies. Of these, the approaches made to BAe were the most significant and potentially the most damaging to a long term European challenge. Despite their established links with the Airbus and other European civil projects, the British were still on the fringes of a full commitment to Europe, and consequently vulnerable to American offers. American initiatives were also helped by the existence of a strong preference in some industrial and governmental quarters for a transatlantic programme. In particular, Rolls Royce and British Airways were determined advocates of transatlantic cooperation.

In the light of the poor commercial record of the British civil aircraft industry, many politicians and officials, especially those in the Treasury, were very receptive to the idea of cooperating with the Americans. The notion that BAe should join a stronger and more commercially orientated manufacturer was very attractive. Conversely, Airbus Industrie was viewed with some suspicion as a French dominated organisation, with a relaxed attitude towards commercial principles. In particular, a deal involving Boeing had the additional advantage of instilling a degree of coherence in British civil aerospace policy. If BAe did join a Boeing programme, and the most likely was the 757, the interests of BAe, Rolls and British Airways would point in the same direction for the first time in nearly a decade. Boeing made it known that the Rolls RB211-535 would be well suited for its 180 seat 757 project, and British Airways had stated quite firmly that a Rolls engined Boeing was preferable to anything on offer from Airbus Industrie. The pressure on BAe to join the Boeing camp became intense, with Rolls and BA both advertising the benefits of cooperation with the Americans. For example, Ross Stainton, BA's Chief Executive suggested that, 'by far the most important factor in the long term outlook for the British air transport industry, airframe and engine manufacturers and airlines, is the fostering of a trans-Atlantic approach to the building, choosing and operation of new civil aircraft'. [21] To support their argument, Rolls and Boeing published data showing the advantages to Britain in terms of employment and sales if BAe were to join Boeing rather than Airbus Industrie. [22]

Despite its links with the Europeans, BAe was willing to consider very seriously what Seattle had to offer. But BAe was not easily convinced that the advantages of working with Boeing were necessarily that obvious. Boeing was prepared to make available to BAe a substantial share of the 757, and detailed discussions on this basis were held throughout 1977. BAe was offered responsibility for designing and building the wing, and in February 1978, Boeing made a formal proposal. Boeing would lead the programme, setting the tooling and engineering standards, and final assembly would be located in the US. BAe would have a fixed-price contract with provision for inflation, but the price would not be allowed to exceed Boeing's estimates of what it would cost Boeing to do the same work. BAe response was non-committal. As the negotiations proceeded, BAe had found Boeing's proposals 'increasingly less attractive'. They amounted to little more than a subcontract relationship, in which BAe could expect to have little influence over the programme as a whole. Boeing insisted on retaining full control over sales and marketing, and there would be no guarantee that BAe would have a place on future programmes. Many observers felt that BAe was deterred by the strict commercial disciplines demanded by Boeing. BAe, on the other hand, denied any fear of meeting Boeing standards, but argued that it was being asked to undertake the newest and most difficult part of the 757 under extremely stringent financial conditions; 'we felt that Boeing was taking no risk, building its same old fuselage, whereas we were supposed to do all the clever bits and take on all the risk'. [23] A slightly improved offer from Boeing, including flight testing and some final assembly work, failed to allay BAe's doubts about the desirability or the viability of the deal. As the debate unfolded, BAe was increasingly concerned to retain an overall competence as a manufacturer of civil aircraft, and with growing conviction, argued that this would be better assured as a full and equal partner in Airbus Industrie. [24]

Shortly before the formation of BAe, a BAC executive had outlined the dangers facing a British airframe company in an American programme. European cooperation was founded upon a rough equality of competence and status, 'we are able to develop organisations in which there is a good standard of sharing of design responsibility'. In a European programme, British industry was more likely to maintain the basis of a fully comprehensive design and marketing capability. Working with the Americans, on the other hand, would mean collaborating with a stronger and inevitably superior partner; 'we would perhaps on the design side be able and allowed to carry out detailed design of parts of the aircraft and so on, but we would certainly have no vital controlling part on the over-all concept'. [25] This was undoubtedly the case with the Boeing offer, and BAe were distinctly apprehensive about its long-term future tied to the Americans. The situation was further complicated by a proposal from MDD for BAe to cooperate on a number of civil and military projects. MDD was prepared to concede a major part of the design responsibility for a civil project to the British. Although BAe was rather more enthusiastic about this idea than anything Boeing had so far come up with, it was still too vague and contained too many uncertainties for BAe to have much confidence in it as a basis for future development. [26]

With an end to French hopes of obtaining a similar deal with the Americans, time ran out for BAe. Once the French accepted that they had to put their full weight behind Airbus developments, they, and the Germans, would not wait for the United Kingdom to make up its mind about future aircraft projects. Contingency plans were already being made for the Germans to assume responsibility for Airbus wing design and production. So that long lead time items could be ordered, Airbus Industrie wanted a firm decision by May 1978.

The whole matter of which programme, if any, BAe should join, became a highly sensitive political issue. The Prime Minister James Callaghan took personal control of the Cabinet Sub-Committee remitted to consider the issue. The Cabinet was, in fact, divided on the merits of a European or an American strategy. A majority favoured an American project of some description, and most of these expressed a preference for Boeing. More talks between the British Government and Boeing and MDD were held during Callaghan's visit to the US during the summer of 1978. The British Government, however, began to have second thoughts about an American programme. According to Newhouse, Callaghan was dismayed by the kind of contract Boeing wanted and the implications for BAe's future place in the civil aircraft business. As 'Tex' Boullion of Boeing recalled, 'they never trusted me. It was always: "What haven't you told us?" They were worried that we would drain their talent and take control of them. They thought this deal would have made second-class citizens of them'. [27] To be fair, Boeing was being consistent with its long stated policy towards cooperation; it was the market leader, and it had to be the superior partner. But it was not what the British wanted to hear. Similarly, under closer scrutiny, the MDD proposal looked less attractive and simply faded away. Callaghan later described his conversations with Sandford McDonnell, 'Sandy McDonnell put forward an attractive offer, but it was never as hard as Boeing's offer. It was never really put forward as a concrete proposition on which you could build a negotiation.' [28] In short, the Cabinet was increasingly doubtful about the terms of any cooperative venture involving the Americans, and were becoming aware of the industrial and political implications of forcing BAe to work with the Americans.

The Labour Government had nationalised the airframe industry with a pledge that it would allow BAe the freedom to make its own technical and commercial judgements. The Conservative Opposition had poured scorn on this guarantee, noting that as a Labour ex-minister, Lord Beswick was to be BAe's first Chairman, a publicly owned company was always going to be under the influence of a socialist government. As BAe made public its determination to join Airbus Industrie, rejecting both American offers on commercial and industrial grounds, the government faced an increasingly embarrassing political choice. If it had insisted on a transatlantic programme, the government would have been presented with the possibility of a bruising political fight with the newly-nationalised company over its commercial autonomy. It would have been more than likely that the whole BAe Board would have resigned over the issue. On the diplomatic front, the German Chancellor, Helmut Schmidt, suggested that the British Government ought to demonstrate its commitment to Europe by allowing BAe to join Airbus Industrie. Finally, Callaghan was impressed by a study made by Sir Kenneth Berrill, head of the Cabinet's Central Policy Review Staff, which indicated that the commercial and industrial aspects of the Boeing and Airbus offers were much closer than had hitherto been believed. [29]

BAe's case for the European option was finally accepted, and after overcoming some last minute problems, duly became a full member of Airbus Industrie. A serious threat to the unity of European aerospace, perhaps even more serious than the French case, had again retreated. European suspicions about American motives, and the fear of long term technological domination had once more been decisive. The offers from Boeing and MDD were certainly very tempting. But equally the British Government had recognised the importance of protecting the industry's 'seed corn'—its design and marketing capability. If Britain was to play an effective role in any joint venture, either with Europe or with the United States, it was vital that it had something to bargain with. As one Minister argued, 'it would be disastrous for this country if we were to decline into being an assembling nation for other people's designs'. [30] BAe could have changed its orientation; there was certainly genuine concern that the emotional symbolism often associated with the Airbus, could weaken its financial and commercial discipline. Far outweighing these doubts, however, was an undoubted commitment by many at BAe to the Airbus and the principle of European cooperation. The pattern of the preceding decade and a half was hard to break. The substantive industrial linkages which had been created between BAe and Europe, exercised a powerful influence on corporate attitudes. As far as Airbus Industrie's supposed commercial laxity was concerned, BAe was convinced that its presence as a full member would help to instill discipline. The fact of the matter was that BAe, like Aerospatiale, was ultimately far happier with even a flawed European reality than an uncertain promise of American cooperation.

Fokker and international cooperation

In terms of both political and industrial significance, the various attempts to reach collaborative agreements between American companies and the British and the French were clearly the most important of the period. However, the Dutch company Fokker was also involved in similar negotiations, and its place as Europe's third most complete civil airframe industry made it an attractive partner and a possible bridgehead within the European market. In terms of turnover and employment, Fokker was ranked behind British, French and German aerospace companies. However, Fokker had an impressive record of producing technically and commercially sound products. Indeed, even the German industry could not match the Dutch as an independent design and development centre. Nevertheless, Fokker's size, the limitations of its tiny home market and financial base, even though it has had considerable support from the Dutch Government, forced it to think in terms of international collaboration. In the 1950s, Fokker built the highly successful F27 turbo prop airliner, but in order to finance a jet successor, the F28, the Dutch Government insisted that Fokker find a risk-sharing partner. Fokker in fact actively supported the principle of collaboration, and from the early 1960s, had links with the German company VFW. In 1968, the two companies agreed to a full merger as VFW–Fokker. [31]

The merger was welcomed by many European industrialists and politicians as the forerunner of a general movement towards the re-organisation and rationalisation of the European aircraft industry on a trans-national basis. Even the most optimistic did not doubt that this would be a difficult and problematic process, but the idea of going beyond *ad hoc* cooperation attracted an impressive following. [32] Mergers seemed to offer a more direct way to end the duplication and competition between European states, and a more effective means of eradicating national bias in European aerospace policy. There were, however, many technical obstacles to overcome, including the absence of European company law, common tax and fiscal regimes. Political issues, such as national control over key technological and industrial assets, defence interests, and the responsibility for financial support, were likely to prove even more difficult.

Initially, the VFW–Fokker merger seemed to drive straight through these problems, and to offer an alternative to the consortium approach represented by Airbus Industrie. In effect, VFW–Fokker was a transnational holding company registered under German law. Its headquarters was located in the geographically 'neutral' city of Dusseldorf, roughly equidistant between Fokker's Amsterdam headquarters and the main VFW factory at Bremen. On the face of it, the two companies had complementary capabilities and interests, with established working links on two civil programmes, the F28 and the German designed VFW 614. Both were involved in a variety of other European projects, and together might expect to carry greater weight in future European projects. It was hoped that the two companies would quickly integrate their entire design, production and marketing operations. As Otto Protsch, VFW–Fokker's Financial Director put it, 'in a few years, the company will be so integrated that the question of who is Dutch or German will be irrelevant'. [33] However, progress towards full integration was slow. In particular, problems were encountered in transferring work from one national base to another, especially where military contracts were concerned. [34]

These issues might have been resolved but for the growing problems with the VFW 614. The original design and development schedule was hopelessly over-optimistic, and was made more complicated by the choice of the Rolls–Snecma M45 engine. The M45 was originally designed for the ill-fated Anglo-French Variable Geometry combat aircraft. Following its cancellation, a civil M45 was to be developed as a collaborative programme paid for in part by the British and French, but mainly by the German Government. The 614 was thus dependent upon the simultaneous development of a new engine, an unusual, and particularly risky approach for any civil aircraft programme. Both the airframe and the engine suffered major setbacks. The engine was affected by the Rolls bankruptcy and the defection of Snecma, and costs escalated.

In 1972, the sole VFW 614 prototype crashed during a test flight, further delaying development and adding to already alarmingly high costs.

VFW–Fokker had estimated that 250 aircraft would be sold, but by the end of 1975, only ten had been ordered. As Fokker was responsible for only 20% of the programme, most of the cost was carried by the German Federal and Lände Governments. Up to 1975, the 614 had cost DM361 million, with more than DM550 million authorised for production. As a result of these expensive problems, the 614 attracted considerable domestic criticism as 'Germany's little Concorde'. The 1975 Grüner Report concluded that 'as things stood it would hardly be possible to make a profit if the programme continues...'. On the whole, the Federal German Government's high financial commitment for this individual project is disquieting from a purely economic and budgetary point of view'. Grüner, however, took account of other factors, especially the effects of cancellation on VFW–Fokker. These were sufficiently important to 'justify giving the VFW 614 another chance'. [35]

The 614's problems, and their effect on the German half of the partnership, did not help the process of integration. Although Grüner recognised the great efforts being made by Fokker's experienced sales staff to promote the aircraft, there was a growing feeling in Germany that the Dutch wanted its cancellation. Fokker had never been entirely happy about the 614. Initially it had been a relatively cheap price to pay for the merger, but many at Fokker felt that the 614 overlapped the F28, and that only its importance to Germany was protecting it. Employment pressures in both Holland and North Germany increased national sensitivities. 1100 VFW employees were laid off because of the 614's problems, while at the same time Fokker was training Dutch personnel to work on the multinational F16 fighter programme. The North German press was increasingly hostile towards the Dutch, and one Bremen politician even filed suit against Fokker claiming that money allocated to the 614 was being diverted into the F28 sales campaign. [36]

In September 1977, as a last attempt to salvage the project, full responsibility for the 614 was handed back to Bremen. Although the aircraft would still have the support of the VFW–Fokker marketing team, it was hoped that the transfer would encourage the German Government to bail out the programme. In the event, the Federal Government decided that cancellation was the only sensible option. In November, VFW announced the end of the 614, writing off DM1150 million, most of which had come from public sources. [37] The decision exacerbated growing speculation about the future of VFW–Fokker. It had been clear for some time that the German Government wanted VFW to merge with MBB, thereby continuing the rationalisation of German aerospace. The presence of Fokker in an enlarged MBB would clearly have presented serious political problems. [38] For their part, the Dutch did not want to become an adjunct to the German aircraft industry. Moreover, Fokker was increasingly optimistic about its future as an independent company with a growing defence sector and a bouyant civil business based on the F27 and the F28.

After a lingering process of separation, VFW–Fokker was finally dissolved in February 1980. As John Ramsden of *Flight International* wrote, 'it will be a long time before another cross-border marriage between aerospace nationals takes place'. He believed that while the 'in-house' competition between the 614 and Fokker's existing airliners had precipitated the split, more fundamentally he asked, 'how can sovereign national Parliaments give priority to spending their taxpayers' money on work for foreign factories when their own workers' jobs are at risk? This is an over-simplification, but sovereignty was, and remains the "bottom line" in cooperation'. [39] For Fokker, the joint company's collapse came as something of a relief. Earlier idealism about its place in a Europe-wide industry had been replaced by a more confident assertion of Dutch independence. Its own products were selling well enough, and Fokker was interested in developing an aircraft substantially larger than its existing range. Nevertheless, it was still necessary for Fokker to find collaborative partners, or at least other firms willing to assume some of the risk of producing a new range of designs.

Fokker was, however, in an awkward position given its determination to maintain an independent design, production and commercial capacity. On the one hand, Fokker did not see a satisfactory future tied to the Airbus consortium. Its position was similar to both the French and the British industries cooperating with Boeing. Although Fokker was an associate member of Airbus Industrie, subcontracting parts of both the A300 and the A310, it believed that its hard won expertise in civil aerospace would be undermined by full membership of Airbus Industrie. On the other, Fokker's ambitions were qualified by its evident need to find foreign partners if it was to remain in the civil aircraft business. Fokker, therefore, began to look for other partners in Europe, the US and the Far East.

Cooperation with Boeing was initially considered, but as Boeing had no immediate interest in launching another, smaller aircraft, Fokker did not want just subcontract work on the 757 and 767. MDD, still looking for partners to develop a new 150–80 seat airliner, seemed the more likely partner. It too faced difficulties raising money for new airliners, a problem made worse by the DC10's difficulties and internal doubts about civil aerospace in general. In May 1981, Fokker and MDD signed a Memorandum of Understanding to develop the MDF–100, an amalgam of Fokker and MDD designs. It was aimed directly at the new 150 seat market in competition with the proposed Airbus A320. To facilitate development, a 'joint venture programme office' was established, which would be staffed equally by Fokker and MDD personnel. This would report to a Supervisory Board, and would again be drawn equally from each company. The management structure would, therefore, be more akin to CFM International than Airbus Industrie, drawing directly upon the skills and facilities of the parent companies, with each individually responsible for its own costs. The arrangement would also allow both to compete in other market sectors. MDD and Fokker hoped that others, particularly the Japanese, would be drawn into the programme. Fokker expressed considerable satisfaction with the MDD agreement. It seemed to have none of the drawbacks which Fokker saw in cooperating either with Boeing or with Airbus Industrie. As Fokker's Chairman, Frans Swartouw put it, 'Airbus would have exposed us to political pressure and Boeing to commercial pressure.' [40] Fokker also believed that many European airlines were keen to see a 'European' alternative to Airbus. At the same time, MDD would provide access to the US market, and experience of managing a big programme and selling large airliners. In return, MDD got a well-financed and technically-sound partner.

At first, matters proceeded well enough, but like other transatlantic collaborative proposals, as the two companies moved on to more detailed issues, the joint programme simply evaporated. In March 1982, the MDF100 agreement was abruptly terminated. Publicly, the depressed state of the market, the lack of a suitable new engine and the failure of key airlines to make up their minds about the need for a new 150-seat airliner were given as the main reasons for the breakdown. American analysts, however, had already suggested that the liaison between MDD and Fokker was unbalanced and fragile. Despite the aid it got from the Dutch Government, Fokker was still a very small company compared to MDD, and its firm adherence to equality may well have been an irritant to the larger American company. There were also signs that Fokker and MDD had diverging technical and commercial ideas, with MDD wanting to build a larger aircraft than Fokker, and to delay its launch beyond 1987 so as not to obstruct the sale of the DC9-Super 80. [41] Indeed, both Boeing and MDD were increasingly committed to the idea of using derivative aircraft to meet the 150 seat requirement, and felt less need to spend considerable sums on new programmes. This approach obviated the need, in the short term at least, to find collaborative partners, and the industrial and commercial compromises that this would inevitably entail.

Fokker, of course, no longer had a new product to follow the F28. Nor was its own financial and industrial predicament any better. However, Fokker held a vital place in the Dutch economy and the Dutch Government realised that it would have to protect a major national asset. In 1984, with the Dutch Government providing $276 million of the $550 million needed for development, Fokker launched the F100. Shorts of Northern Ireland and MBB later joined the programme as risk-sharing partners. Fokker has also signed up British, Canadian, Belgian, French, German and Japanese companies to subcontract work on the smaller F50, a replacement for the F27. These two projects represent the limits of what Fokker can safely develop outside a more extensive collaborative framework. But as such they ensure an independent Dutch presence in civil aerospace. Fokker's experience suggests that a relatively small company can co-exist with Airbus and the Americans if it chooses its product and market carefully. Indeed, the success of the F100 in winning US orders has been sufficient to deter Boeing from launching a 100–110 seat version of the 737. Nevertheless, although the small airliner market is expanding, particularly in the de-regulated United States, it is a highly competitive business, and Fokker will face a strong challenge from Europe and 'Third World' producers entering the civil market for the first time.

The collapse of the MDD–Fokker project effectively marked the end of the 'American offensive' in Europe. Although both Boeing and MDD would continue the policy of spreading the risks of development and broadening the commercial appeal of their products through multinational subcontracting, in the short term, they would have less interest in full scale collaborative agreements. The apparent success of the 'derivative' approach enabled them both to cut the cost of launching aircraft to meet the 150 seat requirement, and to postpone investment in new projects. In the longer term, more radical technical designs could be evaluated and, when the time was ripe, Boeing and MDD could then look for risk-sharing partners. However, if for the moment American interest in partnership with Europe had waned, the prospects of cooperating with the Japanese were much more inviting.

Cooperation with the Japanese

Since the early 1970s, the Japanese aircraft industry, heavily backed by the state, has looked for cooperative agreements with established airframe companies. Initially, the Japanese sought to expand their basic airframe capability through subcontracting, but increasingly looked for full risk-sharing partnerships which would lead to technology transfer, and experience of marketing and commercial exploitation. Always cautious about the risks inherent in civil aerospace, the Japanese identified the market leader Boeing as the safest route into civil aircraft development. However, in the light of Boeing's reluctance to transfer technology or to relax its control over joint programmes, other options, especially cooperation with Airbus Industrie, were considered.

Japan's first foray into collaboration was not entirely satisfactory. They had hoped for a major role in Boeing's 767 programme, but in the event, they only got 15% of the aircraft on simple subcontract basis. While the Japanese learnt a lot about some aspects of civil airframe construction and marketing, it did not generate the kind of technology transfer they had hoped for. Indeed, as one Japanese newspaper noted, 'as it is no longer a joint venture, Japan's influence over the programme is very small, and it will be very difficult for Japan to obtain technological information arising out of the programme that it originally planned to acquire in return for the billions of Yen invested'. [42] Nevertheless, working on the 767 was a major breakthrough for the Japanese aircraft industry, demonstrating its quality and reliability as a partner. Japanese companies also won 70% of non-American equipment contracts on the 767, underlining the potential for wider benefits to the Japanese economy through investment in civil aerospace.

The 767 experience encouraged the Japanese to seek a more substantial role in subsequent programmes, and to demand more than a subcontractor status. The Japanese naturally aimed to 'pick a winner', and consequently, market trends were 'watched closely for any clues as to the most appropriate partner to select in the international development project'. [43] This meant that negotiations with the Japanese could be a trying experience for both European and American companies. [44] Between 1978 and 1984, the Japanese considered a number of collaborative proposals, but the most serious contenders were Airbus Industrie and Boeing. MDD's financial weakness, and doubts about its continuing interest in civil aerospace, made a poor impression on the cautious Japanese. By the same token, Fokker's size and its limited experience of building large airliners ruled out a link with the Dutch. Indeed, the Japanese made it clear that they were not interested in major risk-sharing agreements with individual European companies. [45]

Airbus Industrie's proposals were based on Japanese participation in the A320. With Japanese companies sharing the development of the V2500 engine, a V2500 powered A320 was attractive to the Japanese. But as any new 150 seater would offer both the CFM56 and the V2500, this was not an overriding factor. Similarly, while Airbus Industrie recognised that it was better to have the Japanese working with them rather than with the Americans, the Europeans were particularly sensitive about future competition from Japan. Airbus Industrie wanted the Japanese to become a full member of the consortium, with its attendant obligations and limitations, especially on taking part in competing projects. Airbus Industrie had no intention of handing over its hard won technology without guarantees from the Japanese that they would have a long term commitment to the Airbus programme. These conditions were too wide-ranging and restrictive to be acceptable to the Japanese. As a result, negotiations between Airbus Industrie and the Japanese fizzled out in the early 1980s. [46]

The possibility of Japanese participation in the Airbus was always likely to be problematic. The Japanese were inclined to see Airbus as the weaker challenger, and for political as well as commercial reasons, the Japanese aerospace industry generally tended to prefer links with American companies. In particular, as American pressure to redress the imbalance in trade between Japan and the United States began to grow, purchases of American aircraft and membership of US programmes would simultaneously satisfy a variety of policy goals. [47] The Japanese were especially attracted by Boeing's tentative 150 seat proposal, the 7-7. Boeing's strong views about technical and commercial control and project leadership were such that it showed an initial reluctance to concede a genuine partnership to the Japanese. The Japanese, therefore, gave some thought to joining the MDD–Fokker MDF100 project. MDD and Fokker certainly offered better terms as far as programme decision-making was concerned, but the collapse of the MDD–Fokker venture left Boeing as the only viable alternative.

In the event, discussions between Boeing and the Japanese were continually interrupted by the uncertainty surrounding the prospects for a new 150-seat airliner development. Neither MITI, nor its industrial proteges were willing to undertake a massive investment while market conditions were so fluid. Equally, Boeing management was unsure about the nature and the timing of launching a new aircraft. From the late 1970s, Boeing always seemed to be on the verge of starting a new 150-seat airliner, the 7-7, to match the A320. However, the cost of undertaking a third new project within five years in such volatile market conditions, led to considerable internal debate. In 1980, Boeing decided that the 'near-term' would be best served by the derivative 737-300. [48] Boeing felt that the cost of a new airliner would not be justified by marginal improvements in direct operating costs. Boeing, and MDD for that matter, were beginning to argue that it would be better to wait for a new generation of engines, such as the V2500, or the still more radical UDF, as well as other new technologies, promising huge improvements in airliner efficiency, before committing themselves to new programmes.

In spite of these difficulties, by January 1982, talks between Boeing and the Japanese appeared to be nearing a successful conclusion. Boeing eased its position on programme control, and was even prepared to consider opening a second production line for the 7-7 in Japan if this was commercially justifiable. Problems did arise over work-sharing and Japan's financial contribution. Boeing wanted the Japanese to take a minimum of 30% of the project, whereas MITI was only prepared to accept 20%. The Japanese believed that Boeing, heavily committed to the 757 and 767, would need partners if it was to compete against the A320, and felt they could hold out for a more favourable deal. However, Boeing was in no hurry to launch the 7-7 while the 737-300 was selling well, and the company's overall position was strong enough to hold its ground. The cancellation of the MDF100 also strengthened Boeing's hand. In September 1982, Boeing's Joe Sutter stated that although an agreement was not imminent, 'there have been good discussions and there is a lot of understanding now'. [49]

In January 1984, Boeing and the Japanese again seemed to be close to a deal. The two sides had compromised on Japan's share of development costs, with the Japanese carrying about 25% of the estimated launch costs of the 7-7. For its part, Boeing appeared to be more willing to involve the Japanese in research, sales and product support activities. On the other hand, the Americans wanted a larger share of the sales revenue, reflecting their greater technological and commercial experience. In March, Boeing finally announced a letter of understanding with the Japanese. Boeing would have 51% of the programme, with the Japanese accepting the principle of 'some sort of weighted revenue-sharing in our favor'. Japanese firms would also participate 'to the degree that is possible, and mutually beneficial, in all aspects of the program—design development, production, sales and support'. The exact nature of Japanese participation still had to be worked out, and would depend upon the 'interest and ability' of Japanese firms. Nevertheless, the Japanese felt that this time they had won some important concessions. In the view of one Japanese executive, 'this time we have a genuine partnership with Boeing'. [50]

Boeing appeared to have learnt some important lessons about collaboration. The letter of understanding seemed to represent a significant softening of Boeing's views about the terms on which it would cooperate with an 'inferior' power. The Japanese had evidently won some important concessions on technological transfer and access to the commercial side of the business. However, Japanese optimism about the programme was short lived. In February 1985, Boeing declared that the market was not right to launch the 7-7. At a stroke, the Japanese were left without an aircraft. Boeing deny that the agreement was more than just an attempt to pre-empt a Japanese–European programme. Nevertheless, as *The Economist* put it, 'behind their polite laughter the men who run the

Japanese aerospace industry are less than happy with Boeing'. Even Boeing had to concede that they had upset Japanese plans; 'we have thrown their world into a bit of a cocked hat. They're in the process of re-evaluating their ideas'. [51] This is rather understating the predicament in which the Japanese found themselves. With few options other than to hang on to what they had got from Boeing, the Japanese had to wait until the Americans decided on what should be built and when.

Early in 1986, Boeing and the Japanese resumed negotiations, this time on a joint programme to develop the 7J7, Boeing's prop-fan design (the J signified no more than the initials of two Boeing executives). At a total cost estimated to be in the region of $4000 million, for all Boeing's recent profitability, developing this aircraft represented a substantial risk. Japan's low interest rate policy was also particularly attractive for any long term, capital intensive project. It was, therefore, in Boeing's interests to pick up the threads of their earlier agreement with the Japanese. In March 1986, Boeing and the JDAC finally agreed that the Japanese would become full risk-sharing partners in the 7J7. In return for a $500 million contribution to launch costs, JDAC would participate fully in the design, research, development, production and marketing of the aircraft. Investment and revenue will be shared among the partners in proportion to their equity stake. The deal was the most comprehensive agreement involving a prop-fan project, and for Boeing the most egalitarian partnership it had signed with any overseas manufacturer. [52] For the Japanese, participation in the 7J7 should lead to a substantial improvement in their aerospace technology. With this agreement, complemented by indigenous research programmes, Japan could be well on the way to becoming a major aerospace nation by the mid-1990s. Boeing, however, is sanguine about Japan's potential competitive challenge, no doubt believing that if Japan does acquire a modest independent capability, it would represent more of a threat to European companies than to itself.

If an alliance with the Japanese is to prove as decisive for the future of the civil aerospace industry as some suggest it will be, Boeing is now well ahead. Airbus Industrie has not given up the chase, and the Europeans are making considerable efforts to involve the Japanese in the A330/340 projects, or to take sub-contract work on the A320. [53] The Japanese might prefer to confine their attention to the Americans simply because they have the better record of making money from airliners, and for wider political reasons. So far, the role, certainly the potential, of the Japanese aircraft industry remains something of an enigma. At the moment, it cannot go far on its own, but if MITI's objectives were fulfilled the Japanese aircraft industry would be a very powerful force indeed. Even so, through cooperating now, Boeing might well have forged a vital link in preparation for the commercial battles of the next century.

Competition and cooperation in the airframe industry

To suggest that there was an 'American offensive' in the 1970s, deliberately planned to undermine the emerging solidarity of a European civil aircraft industry, would be something of an exaggeration. Boeing and MDD were undoubtedly serious about forging links with a major European company. Their actions cannot be explained simply as an elaborate plot to sow dissent in European civil aerospace. First and foremost, the Americans wanted to ease the burden of developing new aircraft and to protect a market position in Europe. Nevertheless, if the Americans had formed a partnership with companies such as Aerospatiale, BAe, or even Fokker, the solidarity of European civil aerospace would have been seriously threatened. If it was a deliberate ploy, the American approach, with its insensitivity to European interests, was singularly inept.

Boeing's problem, and only to a slightly less extent MDD's, was that its determination to maintain control over the key aspects of development and commercial exploitation effectively deterred BAe, Aerospatiale and Fokker. They were not technological aspirants, such as the Japanese, or with limited industrial ambitions, like the Italians, but possessors of a proud and valuable independent capability. Neither American company was willing to accept the compromises needed to satisfy their legitimate long-term interests. Boeing's attitude was understandable. It was determined to protect its commercial and technical integrity, and felt that its very success gave it the right to determine the terms on which cooperation would be based; 'why should the 90% of the world's commercial production be dominated by the 10%?' [54] Boeing also distrusted the close links which European companies had with their governments. Boeing executives argued that there were crucial motivational differences between private companies and those which depended upon the state. As one Boeing executive put it, 'the Europeans do things differently from us. They talk about collaboration first, then they decide on a product'. [55] Foreign firms and their governments, it was said, were unduly influenced by political, social and industrial objectives. Collaboration also entailed additional administrative and decision-making burdens which had to be balanced against the advantages of risk-sharing, market access and the better export-financing supposedly available to foreign companies. Boeing for one, would not take a 'penalty to go foreign', and resisted agreements which would complicate decision-making efficiency 'just to accommodate international partners'. [56] Although MDD was less openly dismissive of collaboration, in practice, it too wanted a predominant voice in key decisions. If Boeing, or MDD, had wanted to make real headway in either Britain or France, they would have been obliged to make more concessions to European fears of becoming technological and industrial dependents of the United States.

If cooperation between the Europeans and the Americans produced little in the way of genuine partnership, cooperation with the Japanese airframe industry also had its problems. The lure of Japanese development capital, and its powerful position in the civil market, clearly offered many attractions. But the spectre of other industrial sectors lost to Japan hovers in the background. Relations between the Japanese and the established aerospace manufacturers will, perhaps, always be fraught with some suspicion. Boeing evidently believes that in civil aerospace, a technologically and commercially dominant company can force the Japanese to its way of thinking, and can dictate the terms of cooperation. Others are not so sure, and fear the dynamism of Japanese industrial ambition. The future may show that for once the threat of a Japanese industrial offensive may have been more chimerical than real. Nevertheless, no one should underestimate a MITI backed campaign to develop a technological capability and to achieve a commercial break through.

Although the battle between Airbus and the Americans may capture headlines, the civil aircraft industry is a global phenomenon. The importance of American participation in the Airbus, albeit in the form of equipment and engine supply has already been noted. By the same token, the Italians have a large stake in both MDD and Boeing aircraft. Boeing has always advertised the international spread of its subcontractors, particularly when it has been trying to sell its aircraft, or partnerships in its programmes, to other countries. Airbus Industrie lays similar claim to an international network of suppliers outside the four member states. The production of smaller airliners and commuter aircraft is also conducted on an international basis, with cooperation within Europe, or between European and American firms. [57] But the sometimes bitter struggle between Airbus, MDD and Boeing in the large airliner markets is undoubtedly a dominant and politically sensitive issue. The commercial stakes are high, but, perhaps of equal importance to Europe, the question is also one of technological independence and long term economic vitality.

The Americans retain a considerable advantage in dealing with potential partners, whether in Europe or in the Far East. Despite the rising costs of launching new aircraft, the American airframe manufacturers, especially Boeing, can fall back on their own resources if necessary. The Europeans do not have the same room for manoeuvre in any category above the 100–20 seat airliner. For European companies, and for other aspiring producers, cooperation has become the only way to ensure survival in the large civil airliner business. Subcontracting to Boeing and MDD is an option, and has proved an attractive route for the smaller and less ambitious manufacturers, but the interests and ambitions of the major European companies, as well as the Japanese, demand something more than 'metal-bashing' to the Americans. The commitment to the high value-added end of civil aerospace technology, with implications for equipment suppliers and other related high technology industries, is deeply embedded in the established aerospace countries and has been embraced by Japan. Without access to a complete development and commercial process, scarce and expensively acquired skills will wither, and dependence becomes a real danger.

Notes

1. K. Hayward, *Government and British Civil Aerospace*, op.cit., pp.155–6.
2. *Flight*, 24 October 1974, p.567.
3. Ibid., 9 October 1976, p.1135.
4. J. Newhouse, *The Sporty Game*, op.cit., ch. 3.
5. Ibid., pp.197–8.
6. Ibid., p.198.
7. K. Hayward, op.cit., pp.160–3.
8. *Flight*, 19 June 1975, p.960, 27 November 1976, p.1577.
9. *Interavia*, July 1976, p.609; *AWST*, 19 April 1976, p.24.
10. *Interavia*, July 1976, p.610; September 1976, p.851.
11. *Sunday Times*, 9 May 1976.
12. HoL (305), 1975–76, Q479 & 493; *AWST*, 6 September 1976, p.76. According to one British Executive, 'We formed the Group of Six after the Europlane venture folded. Then the Dassault people got upset and demanded to come in with us, so we made it the Group of Seven. Once in, they told us that the Mercure 200 was the only aircraft they were interested in, and by that time we wanted to keep the French involved so we bent over backwards to evaluate the Mercure 200 fairly along with other possibilities. But the French went their own way with MDD without saying a word to the Group of Seven. The French are a Group of One, which is what they apparently were all the time. We could never bring ourselves to believe that a revitalised European civil air transport industry should tie its future to what we considered a failed commercial venture (Mercure 100), but we compromised for the sake of European unity'.
13. Newhouse, op.cit., p.197.
14. *The Times*, 7 December 1976; *Flight*, 27 November 1976, p.1580.
15. MDD's determination to control all aspects of marketing caused a great deal of resentment. During presentations to potential customers, MDD representatives always outnumbered the French, and never allowed the French to lead discussions.
16. *Flight*, p.1578; *Financial Times*, 13 August 1976; 17 August 1976; 19 August 1976.
17. *Flight*, 27 November 1976, p.157.
18. Ibid., 27 November 1976, p.1578.
19. Ibid., 25 June 1977, p.1784. See also, *Air et Cosmos*, 18 June 1977, p.9. The decision was 'a clear French Government statement to put into action a programme policy built on the Airbus family and a new medium range aircraft'.
20. Newhouse, op.cit., p.196. The Mercure was a superb aircraft to fly, but simply the wrong size at the wrong time. Dassault had entered the civil

business in the late 1960s when the French Plan implied a greater emphasis on civil programmes. Aerospatiale disparaged Dassault's civil pretentions, but also felt the loss of leadership even more. General Mitterrand of Aerospatiale said that he had never been happy with the prospect of an American link, and that the hiatus had cost European civil aerospace a full year of fruitless aggravation. See *Flight*, 7 May 1977, p.1234.

21. *Flight*, 22 April 1978, p.1087.
22. Hayward, op.cit., pp.174–82.
23. Newhouse, op.cit., p.202.
24. Hayward, op.cit., pp.177–8.
25. HoL (305), 1975–76, Q476.
26. *AWST*, 27 February 1978.
27. Newhouse, op.cit., p.208.
28. Ibid., p.210.
29. Hayward, op.cit., pp.180–2. See remarks by Gerald Kaufman, British Industry Minister, 'we are not going into cooperation for prestige but to make a profit and to provide secure employment', *Flight*, 16 October 1976, p.1176.
30. See Kaufman's evidence to HoL (305), 1976–77, Qs 598, 611, 613 & 631.
31. VFW also had an interest in civil aerospace, and had launched the 40 seat twin-jet VFW 614 in the mid 1960s. The VFW 614 started out as the first indigenous West German civil airliner, but as a condition of receiving aid from the German Government, VFW was obliged to seek risk-sharing partners. In any case, VFW's own limited design experience and production capacity made collaboration a logical step. As VFW was already collaborating with Fokker on the Fokker F28 twin-jet, an expansion of the existing partnership was equally sensible. Both companies were prepared to consider more than just a limited extension of existing arrangements. Fokker had long advocated cross-frontier integration in the European aircraft industry. Fokker recognised that it needed to broaden its financial and productive base, having reached the limits of what it could achieve within Holland. VFW, on the other hand, had fewer constructive reasons for a merger, but rationalisation of the German aircraft industry had left it somewhat exposed, and VFW had no objection to a more ambitious form of international cooperation. Neither Government opposed the principle of a trans-national merger.
32. Hayward, pp.155–6.
33. *Die Ziet*, 20 April 1977.
34. *Management Today*, Oct. 1974; *Die Ziet*, 20 April 1977; *Flight*, 23 May 1974, p.661; 31 December 1977. The untimely death of Frederik

Diepen, Fokker's managing director, and a fervent advocate of transnational mergers, was another set back to the smooth integration of the two companies.

35. Grüner Report, p.34. For further details of the M45H engine problems, see Hayward, pp.199–200.
36. German fears may well have had some substance. According to an internal memorandum quoted in an article in a German newspaper, Fokker's marketing chief, Alan Buley wrote that 'I have only one task: to sell the F28', *Die Zeit*, 20 April 1977.
37. *Flight*, 12 November 1977, p.1409.
38. *Flight*, 23 February 1980, p.534. In 1977, a report commissioned by the Federal Economics Ministry, the Deutsche Bank had recommended an immediate merger of VFW with MBB. Indeed, the possibility of VFW-Fokker merging with another German company was discussed in the Grüner report, and the Dutch link was even then considered to be something of an inconvenience.
39. *Flight*, 1 March 1980.
40. MDD took great care to reinforce its commitment to a new aircraft, and confirmed that it did have the funds to finance its share of development. Fokker's requirement for two production lines, somewhat extravagant for such an aircraft, was not viewed as being a significant issue. The two lines ensured greater employment in Holland, as well as reinforcing Dutch equality with MDD. By the same token, it was politically and commercially expedient to retain an American production facility. *Flight*, 13 June 1981, p.1815; *The Economist*, 1 October 1977; 9 May 1981, p.90.
41. *Fortune*, 8 March 1982.
42. *Herald Tribune*, 8 October 1979.
43. MITI Report, see *Flight*, 20 November 1982, pp.1521–3.
44. One participant in Airbus Industrie's negotiations with the Japanese said that the Japanese presented 80 questions in one session, and would not, or could not answer the six asked in return.
45. This did not exclude, however, ad hoc commercial agreements. In 1984, Fuji Heavy Industries took on a sub contract with Fokker to build parts of the F50.
46. *AWST*, 30 May 1983, p.319; *Flight*, 20 November 1982; 23 May 1981, p.1529; *Fortune*, 21 March 1983.
47. *Fortune*, 21 March 1983.
48. *Flight*, 13 September 1983, p.1069.
49. *The Economist*, 8 August 1981, pp.62–3; *Flight*, 18 September 1982, p.836.
50. *Herald Tribune*, 20 March 1984. *AWST*, 16 May 1983; 23 January 1984, p.30; *Financial Times*, 20 September 1983; *Flight*, 4 February

1984, p.310. Such was the uncertainty that one important member of the Japanese consortium was reported to be having doubts about the whole venture. *Flight*, 14 January 1984, p.52.

51. *The Economist*, 'Civil Aircraft Supplement', 1 June 1985, p.22.
52. *Flight*, 15 March 1986, p.4.
53. *Flight*, 11 August 1984, p.4.
54. *Interavia*, Sept. 1976, p.851; Newhouse, op.cit., p.138; *Iron Age*, 14 April 1982, p.33. If Boeing was to make concessions, as, for example to the Japanese, they would be based on money. As Joe Sutter put it, 'Boeing is willing to release more action to partners, on the understanding that they provide the money', *Interavia*, February 1982, p.138.
55. *AWST*, 13 September 1976, p.21.
56. Tom Bacher, *AWST*, 30 May 1983, p.227.
57. *Interavia*, Nov. 1981, p.1174. A particularly interesting example is the BAe 146, where Avco is a risk sharing partner developing the wing, and supplies the engine under a normal contract.

4 Cooperation and Competition in Aero-Engines

Collaboration and competition in the engine industry

Before the advent of the turbo fan in the mid-1960s, with few exceptions an airline would buy an aircraft and its engine as a unit. [1] For the engine manufacturer, the main commercial task was to convince the airframe company to choose its product. The pattern is now for the airline to select the engine after ordering an aircraft. Although there is still some advantage in being selected as the 'launch engine' by an airframe manufacturer, nearly every airliner sale is now followed by an intense battle to supply engines. In their efforts to defend a position, or to break into a market, a complex Dutch auction involving price, performance and other inducements has often occurred between engine companies. For example in 1980, the Rolls Royce RB211-535, 'launch engine' for the Boeing 757, faced a powerful response from the P & W JT10D (PW2037). Two vital orders were at stake, including what was then the largest ever single order for turbo fans. The competition hinged upon P & W's willingness to underwrite the predicted capabilities of a new design against the known performance of Rolls' RB211-535. P & W's tactics led Rolls' Chairman, Lord MacFadzean, to claim that the possibility of P & W 'making money out of these sales is…zero. They are going to make a heavy loss on it. But they would defend that on the basis that they came late and had to break into the market…'. [2]

In a similar case, GE saw off a challenge from P & W which threatened a key sale to an A310 operator, Air France. GE, with its French partner Snecma, had a solid command of the Airbus market. Again, P & W made a generous offer in order to undermine GE's position. Air France accepted P & W's proposal, and asked permission from the French Government to buy the P & W engine.

Although the result was perhaps hardly surprising given the 'political alliance' of GE and Snecma, GE still had to work hard to better P & W. GE undercut P & W's lowest bid, offered improved terms on an overhaul contract, and gave the airline the equivalent of a 'most favoured nation' status, guaranteeing it the terms of any better deal struck with other customers. Finally, GE agreed to set up a joint GE/Snecma parts factory in France, contributing between $30 and $40 million of its cost. [3]

The combination of this highly competitive environment and the remorseless increase in launch costs has driven the major engine manufacturers into international collaboration. The development of a large civil aero-engine is a more complex and protracted task than building an equivalent airframe. Civil engines have to satisfy extremely high reliability and efficiency criteria, and are usually built to more demanding standards than are military engines. The costs and risks of development are concomitantly high, and have grown with every subsequent generation. A new engine, such as the V2500, will cost over $1.25 billion. To match aircraft 'families', engine manufacturers also need a range of products from under 12 000 lb to over 60 000 lb of thrust. To some extent this can be accommodated by 'up-rating' or 'de-rating' a basic design. Although this is not as expensive as producing an entirely new engine, major derivatives, such as the -535 version of Rolls' RB211, may cost over £250 million. Similarly, continual improvements have to be made in response to the competition, and achieving a 1% enhancement of efficiency costs about $25 million.

The monetary problems associated with developing civil engines have, of course, been affected by the depressed state of airline finances and the price competition between the airframe manufacturers. However, the engine companies claim that they have borne a disproportionate share of this competition. Their investment burden is much the same as the airframe producers', but they receive only a quarter of the revenue from each sale. The current emphasis on twin engined airliners has also depressed the volume of engine sales. Even more galling, the steady improvement in engine reliability and durability has reduced the value of spares and replacement engines, hitherto a very important element in engine profitability. Even so, the price of spares has shown an amazing resilience. One airline has produced figures to show that the average price of spares has run well ahead of general inflation.

Over a 15- to 20-year period, the purchaser will spend three to six times an engine's original costs in buying spares. [4] These rewards provide the incentive to bid each time an airline looks for repeat orders of an aircraft, and accentuates competition between the engine manufacturers. Under these conditions, the 'rule of two', that in any given market, only two products can share sales profitably, holds especially true for civil engines. The logic of rising costs and narrowing markets has forced all the engine companies into joint programmes or 'project sharing' exercises. As Lord McFadzean put it, 'the high cost of developing a new engine…is probably beyond the capacity of any one of the companies…'. [5]

The need for collaboration has been further stimulated by the rate at which new engines are introduced. Not only are they more expensive, but the advent of a new engine is less frequent. With a design cycle of seven years, if a company is to maintain a stable level of output, it has to be involved in as many different projects as possible. Collaboration is the best hope of achieving this goal. By the same token, broader national participation, though not necessarily to the direct advantage of the bigger firms, helps to expand the base market. This is especially important for Rolls and Snecma, as American prejudice against foreign equipment can be very high. Rolls, with its turnover almost entirely related to aero-engines, based on a small domestic market, has been the most vulnerable to these pressures. But even GE and P & W have to be profitable within their parent corporations, and they too have found it desirable to spread the risk of civil programmes. For Snecma, cooperation has been essential in its rise to a fourth place in the world civil engine industry. In the case of even smaller and less experienced companies, such as MTU of Germany, Fiat of Italy, and for the ambitious Japanese, cooperation is the only way into civil engines. Odilo Muehling of MTU summed this up in stark terms, 'for the company the size of MTU collaboration is the only economic way of successfully doing business'. [6]

The pattern of cooperation which has emerged in the engine industry is in marked contrast to the airframe sector. The most significant collaborative programmes, the GE–Snecma CFM56 International, and the Rolls–P & W–Japanese V2500 International Aero Engine consortium, are centred on transatlantic partnerships. The situation is further complicated by overlapping contacts between companies involved in other programmes. MTU is a member of IAE, as well as working with GE and Snecma on the CF6. Even more significantly, GE and Rolls have a profit/programme sharing agreement covering two important engine categories. However, cooperation coexists within a broader competitive framework; membership of the IAE consortium does not lessen the sales battle between the PW2037 and the RB211-535. Although the value of joint ventures is readily conceded, reaching collaborative agreements and making them work when companies are simultaneously competitors, is far from straightforward. Relations between the parties, particularly in respect of technology transfer, can also provide sources of tension and suspicion.

Snecma, GE and CFM International

Snecma's relationship with GE began in 1969, when Snecma undertook to subcontract a version of the GE CF6 for the Airbus. This was followed by an agreement to develop the CFM56, '10 tonne' engine, and an international company, CFM International, was formed to coordinate development and to sell the engine. The CFM 56 has made impressive commercial progress, including a major breakthrough in American civil and military markets. Alongside the Airbus, the development of the CFM represents a major element in French civil aerospace policy, substantially improving French technological capabilities. The origins of the GE–Snecma agreement also highlight a deep, and now perhaps unbridgeable divide in the European civil aerospace industry between Rolls Royce and Snecma.

Origins of the GE–Snecma relationship

In the early 1960s, an Anglo-French approach to civil engines seemed well-favoured. In November 1961, the British firm Bristol Siddeley Engines (BSE) and Snecma signed an agreement to develop the BSE Olympus for the Concorde. BSE and Snecma were 'old friends', with a long history of cooperation. Indeed, BSE's link with the French was an important reason for its selection as the British engine contractor on the SST engine programme.

Neither company was big enough effectively to dominate the other, and unlike their airframe colleagues, the two engine companies quickly achieved a close and effective partnership. BSE, having conceived the basic Olympus design, led development, but Snecma derived considerable technological benefit from the experience. In addition to the relationship with BSE, Snecma had long-standing contacts with P & W, and Snecma's tooling was designed to suit American techniques and standards (P & W also held 10% of Snecma stock and had a seat on its Board). [7] Consequently, when the Airbus programme was launched, BSE and Snecma proposed jointly to build the P & W JT9D under licence. Snecma and BSE regarded this as a 'European' programme, which would give them access to superior American technology, but without the problems that were believed to be associated with a close association with Rolls Royce, the dominant European company. Neither firm, but especially Snecma, wanted to freeze the European civil engine industry at a point where they would be seriously disadvantaged by their comparative weakness.

Rolls opposed the BSE–Snecma proposal, as it implied a direct threat to European markets from its main American rival. It would also have robbed Rolls of what was then an important launch order for its own new turbo-fan design. In letters to the British Government, Rolls challenged the American based programme and questioned the Snecma–BSE combination's competence to build an advanced engine. According to Rolls, the JT9D licence was not a genuine 'European programme', and would give the Americans a commercial foothold in Europe at its expense. Rolls lobbied hard, and successfully, to convince the British Government that it should support its bid to provide the Airbus engine. Rolls also moved unilaterally to remove the threat by taking BSE over in 1966, and subsequently allowing the P & W licence offer to lapse. [8] The new company was, by employment, the world's largest engine company and its creation intensified French apprehension about Rolls' leadership of any European engine programme. In the event, the Rolls RB207 was chosen as the basis for a joint project with Snecma and the German firm, MTU. However, the French, rightly as it turned out, suspected that Rolls assigned a lower priority to the Airbus engine than to its campaign to win an American contract. [9] When the British withdrew from the Airbus in 1968, the decision was largely determined by Rolls' problems in developing the RB211 for Lockheed. Consequently, Snecma and MTU turned to the United States to find a suitable partner.

Rolls and Snecma have different explanations for the breakdown in collaboration. The French were unhappy with the relative size of the two companies. Although Rolls and Snecma achieved a reasonably cordial working relationship on the RB207, the French felt that Rolls was reluctant to make the necessary concessions on technology transfer to form the basis of a genuine partnership. For its part, Rolls believed that Snecma, supported by the French Government, wanted to develop a fully independent civil capability, and that this would erode the commercial discipline necessary for an effective programme. In the final analysis, it seems that Rolls and Snecma at best were uneasy partners, a situation made worse by very poor relations which existed between senior figures in both companies. In retrospect, both companies feel that an all European engine for the Airbus would have deterred prospective customers, and that an American partnership was best for both their interests and the Airbus.

Following the collapse of the RB207 programme, Snecma's initial preference was to re-establish its contacts with P & W, and Snecma pressed Airbus to select the JT9D. The Airbus team, however, preferred the GE CF6 on technical and commercial grounds. The Airbus partners made it clear, however, that if all other factors were equal, then a substantial French and German input to an engine programme would be decisive. As GE was trying hard to attack P & W's domination of the civil engine business, it had fewer inhibitions about cooperating with the Europeans, and as a result offered more generous terms than P & W. The Airbus provided GE with a guaranteed access to the European market, and with the French and the Germans contributing to CF6 development, GE had an opportunity to extend the CF6's range at a much reduced cost to itself. Rather than embarking upon a large-scale re-tooling exercise, Snecma went for parts of the CF6 which could be handled by its existing machinery, worth about 20% of the engine. Snecma was also responsible for final assembly and flight testing of engines for the Airbus, which brought the French share of the joint programme to 25%. MTU took another 10%. [10] For a modest investment of Fr2000 million, the French joined an established, advanced-technology programme with, as it transpired, good prospects for further cooperation with GE.

The CFM International

In the late 1960s, Snecma believed that a medium-thrust turbo-fan would be an ideal engine for a 'JT8 replacement' designed for the next generation of narrow-bodied airliners. Snecma had never built a large civil aero-engine, and had to convince the French Government that the project was commercially viable and that Snecma had the necessary expertise to undertake such a project. Early in 1971, the French Government asked for a detailed technical and commercial study. Although the Government accepted Snecma's case in principle and included the project in future planning for aerospace, the company was directed to find a partner to share the cost of development. Snecma had already recognised that cooperation was necessary; as one senior executive observed, 'we'd have been crazy to do it on our own'. Although Snecma held some tentative discussions with Rolls Royce, the latter's financial problems ruled out a commitment to new projects, and the experience of the RB207 affair was too fresh for either company readily to join forces again. In any case, Snecma believed that an American partner would be commercially more credible. The choice, then, lay between P & W and GE. P & W did not want to undermine the chances of further JT8 sales, and had less interest in an early replacement. GE, on the other hand, was still looking to extend its range of civil engines, but was deterred by the cost of a new engine. Although GE was not entirely convinced that a '10 tonne' (22–26 000 lb) engine was a sound prospect, it was attracted by the idea of sharing the costs with a publicly financed company.

By February 1971, Snecma had obtained draft Memoranda of Understanding from both American firms. After a careful review of the two proposals, Snecma felt that GE offered the best terms on technology transfer and work sharing. In March, the two companies signed a preliminary agreement. The engine, now known as the CFM56, was to be based on a full and equal partnership between the two companies. In November 1971, GE and Snecma agreed to the formation of a joint company, CFM International, to supervise development and sale. For the French, CFM International represented a substantial advance over the CF6 contract, with the promise of Snecma gaining considerable technical and commercial experience through joint development. It was also felt that GE, although possessing the necessary credibility, as second to

P & W, would be less likely to dominate the partnership. [11] The French Government provided Snecma with development funding, and extensive provision for the CFM56 was incorporated in the Sixth Plan.

However, French hopes for access to American technology received a considerable setback when the US Government blocked the transfer of vital GE technology. The CFM56 was designed around the GE F101 military engine's 'hot core', then the most advanced engine in the United States. The 'hot core' is the heart of any jet engine, and the most difficult element in an engine's design. Its use in the CFM56 would have brought a significant saving in development costs. Although the American Government accepted that there would be advantages in allowing GE to cooperate with Snecma, there were serious objections to the degree of technology transfer implied by the preliminary agreement. In August 1972, work stopped on the joint programme, and although both GE and Snecma carried on independently, there was no official contact while the question of technology transfer was resolved. Despite high level representations by the French Government (it was a subject discussed at at least one Presidential summit), the American Government refused to budge. GE suggested that cooperation could be organised so that Snecma would handle a 'sealed core' (a complete unit supplied by GE with any exchange of data on a 'need to know' basis). In addition, Snecma would no longer have any responsibility for systems integration, and the design parameters of a civil F101 'hot core' would be reduced to a level not exceeding those of the CF6. GE further suggested that the US State and Commerce Departments could monitor the exchange of data to ensure compliance with American policy. This proved acceptable to both sides, and work resumed on the joint programme in September 1973. [12]

An international company, CFM International, was incorporated in France to administer the joint programme. Shares in CFM International were divided equally between the two companies, and programme costs, estimated at over $1000 million, were apportioned accordingly. Although GE was to be responsible for systems integration and the development of the 'hot core', both parties would have the same proprietary rights over the remainder of the programme. Use of the F101 core also allowed GE to invest proportionately less in CFM R & D. But if this seemed to give GE an undue

Figure 3 CFM International Organisation
Source: CFM International

CFM International has a small staff composed of experienced personnel assigned from the parent companies. It makes integrated use of Snecma and General Electric resources and capabilities.

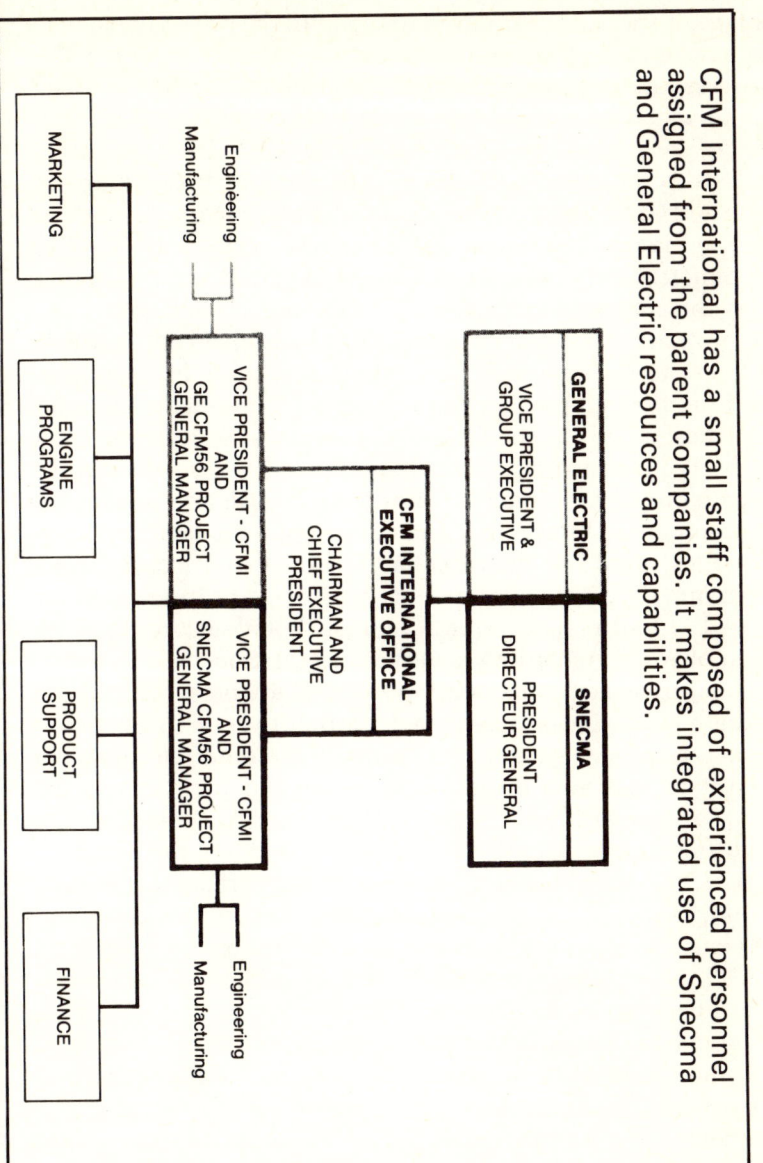

advantage, it was, as one GE executive put it, the price of Snecma's 'introduction to commercial engines'. [13] Snecma accepted this without resentment, for GE's technological expertise and commercial clout was vital to the success of the programme.

The organisation of the CFM consortium was established after a careful study of other cooperative programmes (See Figure 3). Snecma and GE recognised that a central management was needed to oversee the programme, as well as providing a point of contact for customers. On the other hand, GE and Snecma were determined to keep the management system simple and unobtrusive. As a result, initially CFM International only employed 20 full-time executives, to be supplemented when necessary by personnel drawn from the parent companies. Similarly, product and market support has relied on the partners' resources, a decision made easier by the absence of direct competition between the two on other civil projects. Production rates were to be determined by the parent companies. Warranties, and other contractual obligations to third parties, would be shared equally by the two firms. However, as a measure to increase programme efficiency, profits would not be shared but all revenue would be distributed according to the division of work. According to Oliver Fagard, Snecma's Commercial Manager, 'when collaboration takes place on a cost sharing basis there is less incentive for each partner to keep its own costs down because his costs are also borne by the other partner'. The CFM system was also designed to lessen the problems of differential exchange rates. [14] Putting flesh on the agreement was more difficult. It took over 12 months to negotiate the details of work and revenue sharing. In the event, the companies found it impossible to set in advance exact figures, and they accepted that adjustments would have to be made when necessary during the engine's lifetime. Two assembly lines were authorised, an important point as far as the French were concerned, but given the modular nature of engine assembly, this has only added 2% to total programme costs and has reinforced the equal standing of both companies in the consortium.

Although CFM International has a less elaborate organisation than Airbus Industrie, it has acquired a distinct identity of its own. Major programme decisions are taken by the CFM International Advisory Committee consisting of the chief executives of the parent companies, or their deputies. CFM International defines all development and technical targets for the engine and its components, and each company is then responsible for meeting these requirements. A careful division of labour has minimised the number of joint decisions, but the distance between the two centres has led to complex and expensive satellite and other telecommunication links. Consequently, the internal flow of information has not always been effective. On the other hand, the two companies have exploited the time difference between France and the US, effectively doubling the length of a working day. [15]

The CFM56 has a status in French civil aerospace policy second only to that of the Airbus, with commensurate financial and political commitments from the French Government. In 1979, the French Minister of Defence told the National Assembly that the Government's policy was to put Snecma into the ranks of the world's major manufacturers. [16] Governmental support was unquestionably vital in seeing the project through its initial development phase, when the engine's first order arrived only six months before certification. GE had considered abandoning the programme, but had been encouraged by the unwavering commitment of the French Government. Since 1979, over 2280 CFM56s have been ordered by 52 different civil and military customers. This included a $2.7 billion order from the USAF to re-engine its KC135s, in monetary terms, still France's largest military order from the US. The CFM56 was also chosen by United Airlines in its record $3.1 billion order of 737-300s from Boeing, a contract worth over $506 million to Snecma. [17] Between 1981 and 1985, Snecma's commercial sales doubled, and the company emerged as one of France's most important exporters, a success largely attributable to the GE connection. As Oliver Fagard observed, 'it is unlikely that an engine without US content could ever get a contract like the KC135'. [18] For its part, Snecma has always advanced the interests of the consortium within France. When, for example, Air France has expressed an interest in competing products, Snecma has pressed at the highest levels to ensure that Air France bought a GE–Snecma engine. As one executive put it, 'we debug the obstacles to GE to prove that we are a good partner'.

The success of the CFM56, as well as the other engines built in partnership with GE, ensured Snecma's profitability throughout the late 1970s when other French aerospace companies experienced serious financial difficulties. By 1983, Snecma already provided work for 4200 French subcontractors, and continued growth will provide even more stimulus for the French equipment industry. As a direct result of cooperation with GE and, in particular with the need to accommodate the CFM56, Snecma has expanded its plant, improved its facilities and has introduced an extensive programme to increase efficiency through the use of NCMs and other robotic machinery. In April 1983, Snecma and the French Government agreed to a long-term rolling programme to finance the firm's civil production. By the mid-1990s, Snecma aims to have 40% of its work in the civil sector, with between 70–80% of that related to CFM56 development and production. [19]

Snecma has clearly benefited from its relationship with GE, but cooperation has not brought Snecma full autonomy in civil engine technology. [20] Snecma executives, and French officials for that matter, claim that an independent civil aero-engines capability is not, and never was an objective. This has meant that Snecma has had to accept the consequences of dependency on a technically superior partner. Despite the fact that Snecma and GE work closely on the CFM56 and CF6, and have usually agreed on a common strategy, on occasion, the French have had to submit to GE's strictly commercial outlook. In 1980, GE abandoned an up-rated CF6-32 in the face of competition from the P & W 2037 and the RB211-534 despite the fact that the French were prepared to finance much of its development. Similarly, GE baulked at the cost of developing the CFM56-2000, as a more direct competitor to the V2500. [21]

The success of the CFM consortium has stemmed from the quality of its product, and from the mutual regard each partner has had for the other's contribution. GE and Snecma have evolved a similar technical and commercial outlook, and the relationship has been helped by the absence of competition between other civil projects. Collaboration on a range of projects, even though organisationally unrelated, has further reinforced the partnership, a point emphasised by René Revaud, Snecma's President during the 1960s and 1970s, and the man largely responsible for the decision to enter the civil market, '...a single project looks fragile, whereas joint production with GE in a range of engines is to rely upon a stable structure'. It facilitated a wider degree of technology exchange and it enabled the partners more readily to balance work shares. [22] Claude Malroux, another Snecma executive, believed that the success of the CFM56 was due to the fact that it was managed not by a consortium in the classic sense, but through a 'bilateral evolutionary cooperation between two engine companies which have convinced themselves that their long term interests and their respective strategies coincided on a given market, and this without sacrificing any part of their independence, creative potential and ability to sell and manufacture separately'. [23]

Snecma's future as a major civil producer also depends on maintaining its partnership with GE. To this end, in May 1985, Snecma, with backing from the French Government, joined GE's GE36, Unducted Fan (UDF) research programme. As one French official put it, 'the time is now ripe to get on the band wagon for new engine technology'. GE's NASA contract and possible military applications have limited the degree to which Snecma can participate in the UDF project, and is more akin to its partnership with GE on the CF6 than on the CFM56. However, the agreement still amounts to 35% of the project, and according to one Snecma executive, 'we know that we can work together in either type of relationship'. [24] The future of the Snecma-GE partnership depends on the continued success of the CFM56 and the more uncertain progress in the UDF programme. Technically, Snecma needs GE substantially more than GE needs the French. However, both companies have proved that cooperation is mutually beneficial, and whatever form it takes, the GE–Snecma axis will be a major element in the international civil aerospace industry for at least the next two decades. From a European perspective it may be a cause for regret that Rolls and Snecma have, so far, found it impossible to cooperate on mainstream civil programmes. Yet both believe that a purely European engine would not have sold as well as a transatlantic programme, nor, as in Rolls' case, would it have been as successful as its own RB211 range. There may yet be scope for a pan-European project, or a wider programme involving both Rolls and Snecma, but for the moment both companies have looked beyond Europe to provide the basis for a commerical and industrial strategy to take them to the end of the century.

International Aero-Engines

The second major cooperative civil engine programme, the International Aero Engines (IAE) V2500 is the result of a fusion between two partnerships, Rolls and a Japanese domestic consortium, JAEC, and one linking P & W, Fiat and MTU. Unlike CFM International, and the relationship between GE and Snecma, its most important members, P & W and Rolls, are not only technological equals, but are also fierce competitors across a range of other civil products. The origins of IAE are to be found in an earlier, unsuccessful attempt at cooperation between Rolls and P & W.

Rolls and P & W, the first attempt

In the early 1970s, both Rolls and P & W realised that the increasing costs of building a range of engines to match all airliner requirements would be prohibitively expensive. Along with GE, the companies appreciated that the 'rule of two' was going to apply with ever more force through the 1970s. There was no doubting their individual technological competence, but even for the market leader, P & W, and especially for Rolls, still recovering from bankruptcy, the future would be increasingly hard. Although P & W had encountered some financial problems in developing the JT9 for the 747, it held a dominant market position. However, under pressure from GE, and in due course from a revitalised Rolls Royce, P & W had to invest considerable sums to update and to expand its range of engines. Cooperation in at least one engine category made economic and commercial sense, therefore, even for P & W. US anti-trust legislation ruled out links with GE and Snecma was tied up with GE. This left Rolls as the only likely partner.

In the aftermath of Rolls' 1971 crisis, some hoped that Rolls would be more prepared to work with its continental colleagues. However, Snecma and Rolls found it impossible to reach agreement on a suitable project. In any case, despite the unhappy experience with the Lockheed contract, Rolls still felt that its future lay in attacking the US market. If collaboration was necessary, Rolls would be happier working with a technological equal across the Atlantic. Rolls held preliminary talks with both GE and P & W, but GE made it clear that it was both 'married to Snecma' and 'aggressively in competition' with Rolls. [25] Although Rolls and P & W had also been fierce competitors since the 1950s, they found that they both had the same product gap—a '10 tonne' engine for the '727 replacement' to match the CFM56. P & W believed that Rolls' technological status would avoid problems with the US Government over security issues which had limited Snecma's access to GE's 'hot core' technology. P & W also brought with it an existing agreement with Fiat and MTU, Rolls' partners in Turbo Union, which was building the RB199 for the Tornado.

Preliminary talks between P & W and Rolls began in 1973. In the interim, P & W ran a prototype '10 tonne' engine, the JT10D and confirmed subcontract agreements with Fiat and MTU. Launch costs for the JT10 were put at $700 million, with the Europeans providing between 10% and 15% of the capital. The integration of another major partner at such an advanced stage of the programme would be difficult but clearly not impossible if the will to cooperate was strong enough. Cooperation seemed to make sense to all concerned; the Rolls connection would help P & W in the European and 'Commonwealth' markets. Equally, Rolls would be assured of a strong presence in an important engine category at a considerable saving in development costs. P & W, of course, would improve sales prospects for a new engine in the US. Both would benefit from reduced competition.

In December 1975, Rolls and P & W announced an interim agreement; the Americans would be responsible for 54% of a joint JT10D programme, Rolls would take 34%, and MTU and Fiat, 12%. Final settlement depended upon a favourable response to Rolls' application for launch aid and anti-trust clearance from the US Justice Department. The Justice Department accepted that the JT10D was too expensive for a single firm, and by offering an alternative to the CFM56, cooperation would broaden rather than limit competition. On the other hand, it refused to allow cross cooperation between P & W and Rolls on the 5000 lb thrust RB401, an arrangement which had been proposed to ease work sharing problems on the JT10D. Nevertheless, Sir Kenneth Keith, Rolls' Chairman, argued that the programme was essential to the company's survival and it would go ahead with the JT10D. In March 1975, the partners agreed to set up a joint company to develop and sell the engine.

There were, however, a number of outstanding issues. First,

Rolls had to obtain launch aid from the British Government. The Treasury was concerned about Rolls' weak financial position and doubted the company's ability to keep costs under control. Officials wanted to provide the aid at yearly intervals. Under these conditions, Rolls would have had difficulty in formulating long-term plans with P & W. Consequently, Rolls pressed for a less restrictive formula, and eventually, the Labour Government, keen to demonstrate that Rolls, as a nationalised company, could function normally in a commercial environment, accepted Rolls' case for a conventional launch aid package. [26] Secondly, Rolls made it clear to P & W that it had the technological and industrial expertise to be treated as a full and equal partner. The two companies agreed that cooperation would be based on an open and comprehensive exchange of technology. With the approval of the State Department in December 1976, the joint programme appeared to be confirmed. As Sir Kenneth Keith argued, 'in future all major commercial engines would be undertaken as trans-Atlantic collaborations', and that the twin axes of P & W–Rolls Royce and GE–Snecma, 'just about tied up' the international civil aero-engine business. [27]

Despite the accord with P & W, Rolls was still interested in developing variants of the RB211 family, one of which would be close to the thrust range proposed for the JT10D. Rolls claimed that the -535, at 32 000 lb, was in a different category from the JT10D and would not be a competitor. In the summer of 1976, Boeing, the main target for both engines, decided it needed more powerful engines for its new aircraft. As a result, a division of opinion began to emerge between P & W and Rolls over the most suitable response. According to Rolls, P & W's upper limit of 29 000 lb for the JT10 was now too restrictive. Rolls felt that the 32 000 lb RB211-535 would be a more cost-effective way of meeting Boeing's requirement. Boeing's decision also aggravated growing differences between Rolls and P & W over development schedules and cost estimates. Rolls felt that the in-service target date of March 1981 for the JT10D was too tight for a new engine. Rolls also believed that P & W's $500 million estimate of programme costs should be nearer $1 billion. Rolls proposed that the joint programme should shift its design focus from the JT10D to the -535. P & W refused to abandon the JT10D, and, as no compromise could be found, the joint programme was officially abandoned on May 8th, 1985. Rolls stated that the parting of the ways had been 'amicable', and that cooperation in the future was not ruled out.

The failure to reach a final agreement was mainly caused by differences of technical and commercial opinion. However, P & W also appears to have been unhappy with the management system, modelled on CFM International, envisaged for the joint programme. Although P & W, with the largest share of the engine, and holding four of the six main executive positions in the consortium, appeared to have a dominant voice in the project, decision-making was based on 'collegial' procedures. According to a P & W executive, while P & W was willing to accept this during the conceptual stage of the programme, 'when you get to the point of making commitments to customers, you need something more'. This seemed to imply that once the programme was underway, P & W wanted to assume a more central, authoritative role in project management. At the time, Rolls were unaware of any objections to joint decision-making from P & W, and would certainly have objected to the Americans assuming such a leading role.

However, even if there were differences about how the programme should be organised, agreement on fundamentals would have helped both parties to find a satisfactory solution. Its absence made further progress very difficult. 'There are two separate decisions to be made,' observed a P & W official, 'you have to decide how to perform the job once an agreement is signed, but before that you have to decide what the job is that you want to collaborate on'. [28] For its part, P & W believed that Rolls, with its RB211-535 ticking away behind the JT10D, was always going to be suspect as a partner; 'when Rolls decided to build a competitive engine they were indicating that they did not want to participate in the JT10D as it stood. Instead of a collaborator, they became a competitor'. [29] Rolls categorically refuted this charge, arguing that P & W started a new round of competition when, in December 1980, it re-launched the JT10D as the PW2037, aimed directly at the Rolls RB211-535.

In this instance, although financial pressure had led both Rolls and P & W to a preliminary agreement, in the event it proved insufficient to overcome considerable differences between Rolls and P & W on basic technical and commercial issues. P & W was committed to the JT10D because, unlike GE or Rolls, it could not 'de-rate' its basic turbo-fan engine to meet the requirements of a new generation of 130 to 180 seat airliners. When Rolls suggested that P & W should abandon the JT10D in favour of the -535, P & W refused to accept dependence upon another company's design to fill this vital requirement. But neither company was prepared to abandon deeply held convictions about the market and the most valid technical response. The experience would seem to confirm a common belief that two strong, independent companies do not easily become partners. Both companies felt there was an alternative to cooperation, and that they could afford at least one more independent, or largely independent engine programme. Economic realities, however, would increasingly limit the extent to which independence could safely be maintained.

Rolls and the Japanese

By the late 1970s, with Rolls' financial position increasingly strained, cooperation, conceded to be generally desirable, was more vital than ever to guarantee Rolls a presence in every main category of engines. As one executive admitted, collaboration had always been important, but it was 'even more justified under recent circumstances'. [30] With the transatlantic option temporarily closed, and with other European companies locked into one American programme or another, Rolls looked to Japan to find a partner for a new 21 000 lb engine design, the RB432. Rolls already had solid links with the Japanese aero-engine industry. Japanese companies had built Rolls military engines under licence, and the possibility of Japanese participation in Rolls early turbo-fan projects had been considered in the mid-1960s. For its part, Rolls had been much impressed by the performance of the Japanese FJR 710/600 turbo fan demonstrator when it had been tested at the UK's National Gas Turbine Establishment in 1971. [31] The Japanese clearly wanted to build up their civil engine industry, and Rolls realised that it would be better to have them as partners rather than as competitors. In August 1979, Rolls and MITI concluded an interim agreement jointly to develop an engine based on the RB432, designated the RJ500. A domestic consortium, the Japanese Aero-Engine Corporation (JAEC) would be responsible for the Japanese share, and would provide half of the £400 million needed to launch the engine. Rolls, therefore, got a well-financed partner, and the Japanese would be 'fully involved for the first time in a technically challenging aerospace project with enormous potential'. [32]

Rolls' view of collaboration was decidedly pragmatic. According to A. R. G. Raeburn, Rolls' Managing Director, the need for cooperation was purely a matter of financial necessity; 'if we had access to unlimited sources of finance we could do without it'. The approach to Japan was based on five factors; existing links through licence production; Japanese determination to break into the market; their extensive experience in high technology; Japanese interest in airframe development, where the Japanese market would ensure that they would exert a strong influence over customers in the Far East; and finally, Japanese readiness to finance a large share of a big programme. Moreover, Rolls was sanguine about the danger of allowing the Japanese access to its technology, 'lying down with the tiger', as one British Member of Parliament put it. 'We talked about that and we do not think it is [a danger]. Rolls Royce has a long experience of collaborating with Japanese partners in various projects and we are satisfied we shall benefit out of that collaboration as much as they will.' [33]

P & W joins the consortium

Although the Anglo-Japanese agreement provided the basis for an advanced design study, neither side ruled out the possibility of including other partners. The Japanese were especially keen to get an American company involved. The Japanese Government was worried at the rising costs of development, and believed that an American input would improve the engine's commercial prospects. Rolls was rather less eager to involve another major partner, but there was a possibility that Rolls might not have been able to go ahead with the RJ500 as it stood. Rolls' Financial Director said that if they could not attract an American partner it would not necessarily mean an end to the RJ500, but 'we would have to look again'. [34] The Conservative Government, determined to limit public expenditure, also wanted a broader financial base to the programme. There was no question of directing Rolls to expand the partnership, nor was any ultimatum attached to the talks concerning launch aid, but Rolls were given the strong impression that the inclusion of an American company would be welcome. [35] Rolls discussed the possibility of involving GE and Snecma in the RJ500, but with GE and Snecma fully committed to the CFM56 and its derivatives, the only alternative was again to consider cooperation with P & W.

With the JT8 nearing the end of its useful life, P & W needed an engine in the RJ500 class. The PW2037, and other projects were straining P & W's financial position, and realistically, it could not develop a suitable engine alone. As Robert Carlson, President of P & W observed, a three way split of the market for a new engine in the 20 000 lb class would amount to 'self-immolation'. P & W already had a separate agreement with Mitsubishi, one of the JAEC group, to develop high pressure turbine technologies, and the Japanese were very eager to get P & W into the programme. There were a number of problems. Although the failure of their earlier joint venture had not left any deep wounds, Rolls and P & W were fierce competitors, particularly for 757 customers. Indeed, during P & W's initial negotiations with Rolls and JAEC, there was a very bitter public exchange of opinion over the respective merits of the -535 and the PW2037. [36] Japan's technological ambitions provided another problem. Rolls had agreed to a reasonably full exchange of technology with JAEC, but P & W, like many other American companies, was rather more cautious about transferring technology to the Japanese. More encouragingly, Rolls and P & W had complementary technological expertise. P & W's work on single-crystal turbine blades was matched by Rolls' research into wide-chord titanium fan blades. Above all, P & W's involvement in the programme, along with Fiat and MTU, would substantially reduce costs to all concerned.

In February 1983, the five companies announced a provisional agreement to produce the V2500, an amalgam of the best features of the RJ500 and P & W's own design studies. Development and commercial exploitation was to be managed by International Aero-Engines (IAE), an international consortium registered in Switzerland. The agreement was to extend over thirty years, the usual lifetime of a civil engine. After some hard bargaining, work-sharing and financial contributions were eventually settled by Rolls Royce and P & W taking 30% each, with the Japanese having 19.9%, and MTU and Fiat, 12.1% and 8% respectively. The Japanese had held out for 27%, but this was resisted by the two smaller European firms, who felt they were being squeezed by the three larger companies. The partners were to be individually responsible for developing their elements of the V2500 under the general direction of IAE. For the first five years, the IAE Board was to be nominated by the parent companies. Rolls and P & W naturally enough, filled the main executive posts. Although there was to be no programme 'leader', Rolls and P & W would, nevertheless, dominate key decisions and could not be overruled by the smaller companies.

The international management team was based at Hartford in New England, and the engineering offices at Derby. Two design teams, one led by Rolls and the other by P & W, were established, with development responsibilities allocated according to their respective expertise. Two assembly lines, one in Britain and one in the United States were authorised, and like CFM International, IAE created a sophisticated satellite and computer network to aid communication and to exchange data. The division of work along natural technological divisions helped to overcome different attitudes to technology transfer, as well as any objections from the US Government. The V2500 has been designed to be built as two self-contained modules, which are exchanged for final assembly, and minimises any exchange of proprietary data. [37]

As an established manufacturer with its own range of advanced civil engines, Rolls was unconcerned about its lack of access to the 'hot core', a P & W responsibility; but the Japanese and the smaller European companies have had to accept limitations to what they will be able to learn about the most sensitive parts of a modern jet engine. Technology transfer between Rolls, P & W and their respective partners prior to the formation of IAE agreement, were governed by separate agreements, but within IAE technology transfer will be on a 'need to know' basis. In the short to medium term, JAEC has gained considerable experience in the business of building and selling civil engines at a much reduced cost, and there are already some signs that Rolls might have conceded more technology than it would have liked under its initial agreements with the Japanese. No doubt both P & W and Rolls will try to limit Japanese access to key technologies, but in the future, the price of Japanese participation may be a more extensive exchange of technology. In any event, thanks to a mixture of collaborative and indigenous programmes, Japan is already well placed to participate in the next generation of engine development.

The IAE agreement had little difficulty obtaining US Justice

Department anti-trust clearance. As a direct competitor to the CFM56, it was deemed that the V2500 would increase choice in a growing market. US export controls were a different matter. At first, Rolls refused to sign the standard US agreement governing high technology exports. Rolls would not accept restrictions on its commercial autonomy and appealed to the British Government to make representations in Washington. In March 1984, Norman Tebbit, the British Secretary of State for Industry, raised the issue during an official visit to Washington, but the US Government confirmed that it would only sanction the use of P & W technology if Rolls accepted the export control law. Rolls was far from happy, for in common with many European companies, it objected to the extra-territorial extension of American legislation. But the US Government refused to make any concessions; no controls no deal. With some reluctance, Rolls accepted the principle of export control, leaving the exact implementation of the law to future negotiations. As one Rolls executive admitted, 'we agreed to enable us to do business with P & W'.

For the present, IAE has got down to the less political, but equally challenging task of making a five nation consortium work. Once the tough initial discussions had been concluded, the programme ran 'exceptionally smoothly', and the first unit was assembled well ahead of schedule. In this respect, the IAE management was able to insist on a tight timetable, guaranteed by each member of the consortium. [38] The potentially sensitive issue of competition between Rolls and P & W on other projects has, so far, been muted. According to Ralph Robbins, Rolls Chairman from November 1984, the contradictions are 'nothing like as great as one would imagine'. IAE, with its separate sales and marketing division, has enabled P & W and Rolls personnel to avoid conflicts of loyalty, a problem which would have inevitably arisen if the two had been directly responsible for V2500 sales. [39] No one underestimates the problem of organising a complex engine programme across two oceans, involving two highly combative firms. So long as the V2500 sells, however, the consortium has a good chance of achieving a high degree of stability.

Although the CFM56 established a commanding lead in the market for derivative airliners, by early 1986, the V2500 had the edge in the competition to win A320 customers (over 137 orders and options, to CFM's 88). Although CFM claim that theirs is the cheaper, proven design, the majority of A320 airlines, including Pan Am and Lufthansa, have gone for the more advanced V2500. [40] In this context, there can be no doubt that the French Government attaches considerable importance to the link between the CF56 and the A320. [41] Throughout all the uncertainties associated with European airframe projects from 1976 onwards, the use of CFM56 engines was a common factor in French policy. Similarly, French support for the A320 was associated with an equivalent commitment to develop a suitable version of the CFM56. Indeed, the aircraft was often described in the French press as *le bi-CFM56*. Following the A320's launch, Airbus Industrie has tried to be 'strictly neutral' in the competition between the CFM56 and the IAE V2500, and has argued that the choice of engines is up to the consumer. Nevertheless, the competition between the CFM56 and the IAE V2500 may yet provide an interesting confrontation between market forces and the requirements of French industrial policy. IAE remain confident, however, that airlines will want to match an advanced technology airliner with the best available engine, and that the V2500 is clearly the most suitable choice for the A320. By the summer of 1985, IAE had sold over $1 billion worth of engines.

The prospects for future collaboration based on the IAE organisation are more problematic. Rolls and P & W have different opinions about the next generation of engines. Rolls is less confident than P & W that the UDF/Prop Fan engine will replace the conventional jet engine as early as the mid-1990s. [42] However, if the introduction of the UDF should prove as imminent as both P & W and GE believe, it could have serious consequences for the V2500. The V2500 would immediately face competitive pressure from a new technology promising substantial improvements in fuel burn and other performance characteristics. The temptation for P & W to concentrate on a UDF engine could become overwhelming, and it would certainly have little incentive to invest in further developments of the V2500.

Rolls and GE

In February 1984, the announcement of a risk and revenue sharing agreement between GE and Rolls Royce completed a triangle of transatlantic relations. The deal, covering the RB211-534E4 and the CF6-80C2, came as a surprise to commentators, as well as to the two firms' partners in other projects. In effect, GE and Rolls had concluded a 'non-aggression pact', each filling a gap in the other's range which would be too expensive to develop independently. Rolls offered access to its RB211-535E4, at 40 000 lb, in return for a share of GE's CF6-80C at 60 000 lb. The deal was the result of quiet diplomacy—'matchmaking of brilliance', according to *The Economist*—over a number of months between Rolls' Chairman, Sir William Duncan, and Brian Rowe, head of GE's aircraft engine division. Neither side admitted to initiating negotiations, but the arrangement was said to have 'emerged', as a sensible solution to a common problem. [43]

Rolls appeared to get more out of the arrangement than GE. Sir William Duncan had been able to exploit the fact that GE did not have an engine suitable for airliners in the 180–200 seat range. This was a more serious deficiency than Rolls' lack of a very large engine. The CF6-80C was also well entrenched in the A300 and A310, important gaps in Rolls' market coverage, and it kept Rolls in touch with the growing thrust requirements of the Boeing 747. Rolls had little hope of launching a new 60 000 lb engine, and tentative plans to up-rate the RB211 had been shelved for financial and commercial reasons. Additionally, the GE link reinforced Rolls' presence in the US, as GE would help Rolls to sell the -535E4 in North America. [44] Despite huge improvements in productivity, and collaboration in the V2500, Rolls had to find more ways of spreading the cost of maintaining a full range of engines. With a lower share in nearly every engine market, Rolls had to carry a proportionately greater R & D burden than its competitors. Rolls' revenues had been severely hit by recession and its credit line to the British Government was stretched to the limit. Rolls had the additional disadvantage of a less extensive defence sector than either of its two American competitors to compensate for the downturn in the civil market. [45]

Both companies, however, seemed happy enough with the arrangement. Each company would pool the revenue earned by its respective engine, which would then be shared in proportion to the contribution made by each to the other's product. This contribution could be in kind, that is to say, a value would be placed on any technical improvement made by one party in the other's engine. Corporate identity and management responsibilities were not affected by the arrangement, and the firms retained full control over their products. Consequently, there was no need at this stage to create a separate company or joint management structures. The arrangement covered 15% of the two engines, rising to 25%, and contained provision for a joint programme to develop a new large engine in the future if both companies concluded that it was commercially viable.

The firms agreed to an exchange of test and research data to increase mutual familiarity with the other's products, but the need for extensive technology transfer was reduced by the retention of individual control over the development and fabrication of critical components. In this way, either company could develop a new component or process which could be fed directly into either programme, without involving a transfer of technology. The deal carefully skirted US anti-trust laws, and the limited degree of technology transfer circumvented other objections from the US Government. The CF6-80C was already the subject of cooperative agreements with Snecma, MTU, Volvo and Fiat, but Rolls and GE confirmed that their arrangement would not affect contracts with other companies. [46] The agreement was described by Sir William Duncan as a 'very important policy decision, a watershed for Rolls Royce'. The GE deal, Rolls' involvement in other collaborative ventures, as well as its independent projects, kept Rolls in virtually every engine market, from 'the smallest in helicopters to the very largest in jumbo jets'. More important, it was 'affordable technology', maximising sales and minimising R & D costs. [47] For GE, Rowe said it would 'help both companies do better by sharing risks'. In short, both companies felt that they stood to benefit by curbing the rising costs of development, by opening new markets, and by expanding sales volume. It would, in Rowe's view 'help get rid of some of the duplication and should help lead to better products'.

Some observers, including the British journal *Flight International*, believed that this agreement went further than the collaboration between GE and Snecma, or, for that matter, Rolls' relationship with P & W in IAE. Even allowing for the limitations on technology transfer, the link between GE and Rolls potentially could lead to a 'real integration of ideas and information on research, technology, manufacture and marketing'. [48] The general reaction from Snecma and P & W was cautious and largely neutral in tone. P & W officials felt that the withdrawal of Rolls from the higher range of engines would help to improve general market conditions for everybody. Snecma said that the agreement did not in any way call into question its arrangements with GE. It was also suggested that cooperation between GE and Rolls might lead to closer relations between the British and French aero-engine industries. In structural terms, the Rolls–GE agreement introduced a third transatlantic axis, and established a tentative, indirect point of contact between Snecma and Rolls.

Competition and cooperation in the engine sector—an overview

The three cooperative agreements, CFM International, IAE, and the Rolls–GE pact, combined with other subcontracts linking Fiat, MTU, the Japanese and other minor engine manufacturers, have led to a complicated pattern of transnational commercial alliances and obligations in the engine industry. Competitors, even bitter rivals, appear able to coexist in other ventures. On the face of it, such a mixture of cooperation and competition would seem to imply considerable potential for conflicts of interest, yet all concerned claim that any tension or difficulty in working together have been contained. Of course, the problem of simultaneous competition and collaboration is largely a matter of accommodating Rolls Royce and its commercial rivals. P & W does not cooperate with either GE or Snecma, and Rolls has few contacts with the French. In general, however, the success of collaboration in the engine sector lies in the widespread recognition that without some degree of rationalisation, making money out of civil engines would become increasingly problematic. Indeed, it is highly probable that at least one major firm would have to give up the struggle.

Both GE and P & W accepted that the risks and costs of engine development were too great to bear with confidence and safety. Similar economic realities led to Rolls' interest in collaboration. There were also other motives at work. For the French, the acquisition of a civil aero-engine capability was incorporated into a broad strategy aimed at improving the scope of its national civil aerospace industry. Support for Snecma, and the CFM consortium in particular, has become a key element in French industrial and technological policy. The success of the CFM56 and Snecma's growth, with consequent benefits to other high technology sectors and the French economy generally, has fully vindicated their efforts. Finally, the Japanese, and certainly the smaller European firms, have accepted international cooperation as the only way of building up a civil engines capability. It remains an open question whether the Japanese will be satisfied with the kind of partnership they have forged in IAE, or whether they are aiming at an independent capability. Like civil aircraft production, there will be problems in making the transition from partnership to independence, but again, no one underestimates either Japanese skill or determination.

There is no doubt that economic and commercial pressures have been the main stimulus for the current transformation in the civil engine business. Few in the aerospace industry, whether in the engine or airframe sector for that matter, have approached collaboration for altruistic motives. Similarly, despite the complex of collaboration that has grown over the last decade or so, the demise of the national firm should not be anticipated. The major engine companies, P & W, Rolls and GE have certainly not committed themselves to permanent and comprehensive cooperative structures in the way that Airbus Industrie has become the focus for European large airliner development. At a symposium of engine manufacturers held in 1984 virtually every contributor made reference to the importance of maintaining an independent production and technological capability. [49] Quite apart from the difficulties that US anti-trust legislation posed for American domestic collaboration, nobody supported the idea of permanent consortia. In the view of J. M. S. Keen, Vice President of IAE, 'The idea that there will eventually be just two large international companies is a gross oversimplification'. The most significant obstacle is that the greater part of civil engine development and production is still based on independent projects, and all three major companies were determined to retain separate identities and expertise in the fields of marketing and sales. Collaboration, especially the emergence of genuine technological partnership, was also constrained by competition and the fear that an innovation could be exploited by a grateful competitor. In this respect, the modular nature of engine development and production has helped technically-dominant partners in joint ventures to control technology transfer.

There was also an indication that few were entirely satisfied with the way that the existing consortia were organised and managed. The main problems identified by the symposium were, *inter alia*; the difficulty of deciding on the optimum size of central staffs; how to share revenue and costs; how much power partners should have to veto decisions; how to balance the needs of the consortium against those of the separate companies; and finally, how to hold down the costs of cooperation itself. There was general agreement that revenue sharing was the only system that made sense; so that financial discipline was maintained by ensuring that development and production costs remained the responsibility of each partner. As far as the costs of collaboration was concerned, these could be minimised with careful management, and by concentrating effort on established technology with fewer unknowns. In the final analysis, however, performance determined success. If a programme went well, the cost of collaboration would be negligible, but if things went badly, collaboration would only make the situation worse. In short, while the costs and risks of engine development ensure that cooperation will remain an essential feature of the industry for the foreseeable future, existing patterns of cooperation, and certainly the current organisation of joint ventures, should not be regarded as permanent features of the international aero-engine industry.

The intensity of competition between P & W and Rolls in at least two market categories certainly raises some doubts about IAE's long term future. Problems will be muted so long as sales of the V2500 remain buoyant, and different views about the technical and commercial opportunities afforded by an early use of UDF technology do not disturb the partnership. On the other hand, CFM International, or at least the GE–Snecma partnership, appears to have made the transition to a new generation of engines. Snecma clearly had the greatest need to maintain its links with GE, and has been willing to revert to a more conventional subcontract relationship in order to keep in touch with the new technology. However, with the current range of CFM56 engines approaching the end of easy and cheap development, any hiatus in the appearance of a new engine could create problems for Snecma. Further development of the conventional turbo fan could provide a fall-back position for CFM, but much will depend upon GE's interest in investing in the CFM56.

Of the three cooperative relationships, perhaps the most interesting is that between Rolls and GE. This is the most complementary of the major partnerships, a relationship between technical equals, with less competitive friction. But GE has not been happy to see Rolls creeping into the CF6-80's market with the development of a 56 000 lb version of the RB211-524, competing for customers for the larger Boeing 747s and, if its launch is confirmed, the MD11. Rolls has argued that this was just natural growth in response to a particular customer; but one GE executive warned that 'Rolls has got to decide whose wedding it is dancing at'. [50] One interesting possibility is that GE could act as a matchmaker between Rolls Royce and Snecma, perhaps to develop a UDF engine. But, with Rolls still evincing doubts about early use of the UDF concept for civil aircraft, any suggestion that either GE and Rolls, or Rolls and Snecma, might jointly develop a prop-fan engine, would be premature. However, despite Rolls' current scepticism about the imminence of the UDF, and its own research into ultra-high bypass engines, it may be forced into a more rapid and comprehensive programme of development. At the moment, Rolls seems rather isolated from current UDF partnerships, and largely dependent on money which may be made available by the British Government to finance future projects.

In general, the dominant pattern of cooperation in the aero-engine industry comprises various combinations of transatlantic relationships. The emergence of this orientation can be explained initially by the failure of the European collaboration. This may have been a disappointment to confirmed 'Europeans' who would have preferred a European engine industry to match the progress made by the aircraft manufacturers. However, both Snecma and Rolls believe that a European engine would have experienced great difficulty in penetrating world, and especially American markets. Equally, the growth of transatlantic cooperation should also be seen as an important political as well as technological counterpoint to the largely competitive relationship that has evolved in the airframe sector. If tension between the United States and its main trading partners does increase over the next decade, a too obvious transatlantic divide in the civil aerospace industry could have serious consequences for all concerned. In this context, the cooperative pattern formed by the engine companies could moderate protectionist sentiment in the United States. It might also serve to help American and European companies contain Japanese ambitions.

Notes

1. Although some choice was always available (for example, BOAC chose Rolls Royce Conways rather than P & W engines for its 707s), up to the advent of the turbo-fan, the airframe manufacturer tended to have the greatest say in the selection of a suitable engine.
2. HoC 389, 1981–82, Q331. P & W agreed to pay the additional fuel costs incurred by its customers while they waited for the PW2037.
3. *Flight*, 23 June 1979, p.2228; 28 July 1979, p.245.
4. *Interavia*, December 1985, p.1343.
5. HoC 389, 1981–82, Q341.
6. *Flight*, 1 June 1985, p.126.
7. Hayward, op.cit., pp.62–3; Lowell, op.cit., p.53.
8. The merger is regarded as playing a significant part in the problems which led to the Rolls collapse.
9. Hayward, op.cit., pp.81–3.
10. Lowell, op.cit., pp.64–5.
11. J. Baranson, *Technology and the Multinationals*, Lexington, 1978, pp.23–9; *Le Monde*, 8 June 1979.
12. Baranson, pp.26–9.
13. *AWST*, 6 June 1977, p.214; *Flight*, 4 December 1975, pp.831–2.
14. *AWST*, 6 June 1977, pp.213–4, 30 May 1983, p.186; *Flight*, 1 June 1985, p.127.
15. Ibid.
16. *Air et Cosmos*, 8 December 1979, p.3.
17. *Financial Times*, 12 November 1985, 13 December 1985.
18. *Flight*, 1 June 1985, p.125.
19. *Air et Cosmos*, 1 September 1979, p.3; 16 April 1983, p.11; *Flight*, 28 May 1983, p.1481.
20. See General J. R. Soissons, Head of the French DCTA, *Flight*, 9 November 1972, p.652.
21. In the event, GE was able to fill this gap in its range through cooperation with Rolls Royce. See pp.00.
22. *Air et Cosmos*, 23 September 1978, p.11.
23. *International Aerospace Consortia Colloquy*, Feb. 1982, WEU Official Report.

24. There is also a possibility that GE and Snecma might cooperate in the development of a military engine for the Rafale fighter. Snecma–GE cooperation on military engines would be rather more difficult to manage. Existing US vetting of technology transfers would be more complex to organise, and France would have to accept stringent controls over exports. It would certainly imply a considerable departure from French post-war defence–technology policy. *Financial Times*, 13 December 1985, 16 December 1985; *AWST*, 25 February 1985, pp.41–3; 27 May 1985, p.20.

25. *Flight*, 7 June 1973, p.862.

26. See Hayward, op.cit., p.169.

27. HoL 305, 1975–76, Qs 528 & 539.

28. *AWST*, 6 June 1977, p.214.

29. Ibid., 16 May 1977, p.17. See also, *Flight*, 14 May 1977, p.1296.

30. Rolls Royce memorandum to HoC 389, 1981–82, p.5. Rolls' problems were exacerbated by Lockheed's decision to end L1011 production, an aircraft where Rolls had a monopoly.

31. A. M. Anderson, *Science and Technology in Japan*, London, 1984, p.219.

32. *Far East Economic Review*, 31 August 1979, p.84.

33. HoC 389, 1981–82, Q151.

34. *Flight*, 27 February 1982, p.570.

35. Ibid., 17 July 1982, p.136.

36. *AWST*, 7 November 1983, p.29.

37. *Interavia*, March 1985, p.289.

38. *Flight*, 14 December 1985, p.19.

39. *AWST*, 21 May 1984, p.53.

40. *Flight*, 21 May 1984, p.1464; *AWST*, 21 May 1984, p.33; *AWST*, 28 May 1984, pp.20–1.

41. *Air et Cosmos*, 1 September 1979, p.3; 16 April 1983, p.11; *Flight*, 1 June 1985, p.127.

42. The principle of the Unducted Fan (UDF), or ultra by-pass engine, aims to combine the virtues of turbo prop engines with those of the advanced turbo-fan engine. With the help of a $200 million contract from NASA, American companies have been investigating the UDF for some time. It promises roughly the same speed as a conventional turbo-fan engine, but with enormous savings in fuel burn, estimated to be some 24% better than the V2500 (a full use of other new technologies may result in a 60% improvement in operating efficiency over existing airliners). There are, however, considerable technical uncertainties associated with its introduction. There are doubts about the safety of the prop fan itself, whether it will need a gear box, and worries that vibration levels will be both unacceptable to passengers

and dangerous to airframe integrity. Its first use may yet be in a military transport.

43. *The Economist*, 11 February 1984, p.70.
44. Ibid.
45. *Financial Times*, 4 February 1984.
46. Ibid.; *AWST*, 13 February 1984, pp.30–1.
47. HoC 491, 1983–84, Q73.
48. *Flight* (ed.), 11 February 1984.
49. *AWST*, 18 June 1984, pp.108–9.
50. *The Economist*, 1 June 1985.

5 Subsidies, Sales and the State

From its inception, the aircraft industry has depended upon the state. Even in the US, where American manufacturers often claim, in contrast to their European counterparts, that their civil projects are not supported by the state 'one is forced to conclude...that without the federal government, there would simply be no aircraft industry....No aspect of the industry, including the commercial sector, could exist without the R & D funds provided by the state or the state's purchases of military equipment. It is no accident, then, that virtually every other nation has an aircraft industry that is either heavily subsidized or indeed owned by the state'. [1] Some Europeans, indeed, regard state aid to civil aerospace as the only way of redressing the advantages of the large, integrated American domestic market and a huge defence budget 'spilling over' into the civil sector. However, the differences which have appeared between Europeans and Americans about the place of the state in civil aerospace reflect deep-seated political perceptions and prejudices, where high principle inevitably intermingles with commercial and industrial self-interest.

State support for civil aerospace development

The present dispute is deeply rooted in the history of post-war civil aerospace development. Although the Americans might legitimately claim that their current civil programmes are free of state aid, Europeans will inevitably point to the past. The link between the Boeing 707 and the military KC135, is still cited by Europeans some twenty years later, though the risks taken by Boeing in launching the original 367-80 prototype, the additional costs of developing the 707, and the subsequent commercially-based launch of the 727 and 737, are sometimes conveniently forgotten. Although Boeing's later domination of the market was mainly the result of private risk, the original boost provided by the KC135, as well as its large jet bomber programme, was of considerable importance in easing its entry into the civil jet market.

In the 1960s, USAF requirements for heavy lift transports and the needs of commercial airliners did begin to diverge, but research contracts for what became the C5A, helped Boeing, MDD, and especially Lockheed, to launch wide-bodied airliners. The history of the Lockheed L1011 provides an unusually direct instance of the link between American civil and military activity. Ernest Fitzgerald, a Pentagon official remitted to curb malpractices in defence contracting, published a detailed and candid account of the L1011–C5A relationship. He showed how Lockheed, with the connivance of the USAF, was able to transfer money authorised for the C5A to aid the ailing L1011. [2] Even this was insufficient to save the L1011, and Lockheed came close to bankruptcy. Lockheed however, obtained a controversial, and hotly contested, Federal loan guarantee for $250 million to bail out the L1011 programme. Lockheed and its supporters argued that the C5A and the L1011 were 'interdependent' and that the firm's collapse would have had serious implications for the health of the American aerospace industry and its subcontractors. As Fitzgerald wryly observed, 'with Lockheed's commercial ventures underwritten by the taxpayer and their military programs safely tucked back into the Pentagon's cost plus womb and hooked up to the Treasury's umbilical cord, the Great Plane Robbery was an accomplished fact'. [3] Moreover, Lockheed's argument for temporary assistance was helped by a precedent set by Douglas' rescue in the mid-1960s. Douglas then faced a crisis induced by the cost of producing the DC8 and DC9. The company made a successful application for an emergency loan under the 1950 Defense Production Act, the so-called 'V' loan system. Both the Lockheed and the Douglas cases were clearly emergencies, with defence contracts at stake, and the loans were ultimately repaid with interest to the US Government. Nevertheless, neither Douglas nor Lockheed had to pay the ultimate price of commercial misjudgement, and, *in extremis*, their survival had depended upon state aid. [4]

P & W and GE also benefited from work done on behalf of the C5A to launch civil turbo-fan engines. They both won development contracts for a military engine which enabled them to establish the broad design parameters of the high by-pass ratio turbo-fan. GE in particular was able to defray a considerable element of the commercial CF6 programme against the engine it built for the C5A. Indeed, GE deliberately chose to stay out of the commercial turbo-fan market 'until the costs of overcoming basic problems common to all high-bypass engines had been largely absorbed by the Pentagon'. [5] Although P & W risked over $1 billion of its own money on launching the JT8 for the Boeing 747, in both cases public and private capitalisation were again intermixed, a fact which Rolls Royce used to bolster its campaign for direct government assistance to launch a comparable engine. [6]

Newhouse rightly observes that the days of military aircraft programmes 'that once cleared the way for derivative commercial transports' are a thing of the past; yet the 'old' relationship still has its uses. In November 1983, with the imminent closure of the DC10 production line, MDD appeared to be abandoning large civil airliner production. However, the USAF ordered a tanker–transport version, the KC10, and MDD was able to keep its DC10 production line open. The 44 tankers ordered by the Air Force, were not, in the words of an MDD executive, 'going to make a lot of money, but it was one hell of a help with overheads'. In May 1984, Federal Express ordered six DC10s, and MDD were able to restart civil production. Moreover, the combination of the Federal Express order and the KC10 could carry MDD 'nicely to the DCXX', a proposed development of the DC10, which has since been tentatively offered to the airlines as the MD11. [7]

The relationship between American military and civil aerospace goes much further than the connection between civil and military R & D, or 'bail out' loans granted to protect defence contractors. In the past, it has included the 'GOCO' system of leased plant and tooling, tax concessions on purchase of domestic products by American airlines, and a helpful tariff regime. US contractors have also been able to earn high rates of return on military programmes which have permitted 'these manufacturers to initiate commercial aircraft projects whose prospective rate of return, adjusted for risk, has been below that which ordinarily would be required'. [8] The scale of defence activity, and the cash-flow it has generated, has also improved the credit rating of aerospace companies in the commercial money markets. MDD's confident return to the civil market in the 1980s was considerably helped by two decades of uninterrupted success as a defence contractor. This, reinforced by good management, enabled MDD to build up large capital reserves and assets which allowed it to offer very generous credit and contract terms to civil customers. As a means of improving the nations' technological standing, or for enhancing civil programmes, defence spending may be of questionable efficiency. However, the US civil aerospace industry is well placed to benefit from both the cash-flow generated by defence contracting and the general advances in technology which are stimulated by military R & D. Commenting on President Reagan's military build-up, *Fortune* observed, 'the military isn't the best R & D game in town, but its the longest-running game—and for now, the one with the biggest pot'. [9]

European companies also have a defence base which might be seen as providing a similar financial cushion and technological stimulus. Boeing point out that in 1983, the combined sales of BAe, MBB and Aerospatiale was $7.2 billion, with $5.5 billion coming from military work. This compared with its own and MDD's government/military sales of $9 billion. [10] Historically, however, the scale of US defence spending has dwarfed European efforts. The British SBAC has noted that American companies have had a more stable defence procurement base, and proportionately, have had to depend far less on exports than have European firms. It should be remembered that the European defence effort is not unified, and the turnover of American companies is much higher than that of individual European companies. The Americans have also been able to derive considerable advantage from a succession of large defence programmes, and generally, could count on longer production runs. [11] In the 1980s, P & W, for example, sold $2.7 billion worth of equipment to the American Government, a sum equal to the total turnover of Rolls Royce. Similarly, European manufacturers have rarely been able to enjoy such direct links between civil and military requirements as, for example, Boeing did with the 707 and the KC135. [12]

Both the United States and European countries have publicly owned and operated R & D establishments, or let contracts for basic research which help to supplement the general research base for their aerospace industries. Here, a distinction between civil and defence work is hard, if not impossible to make. Quantitative data on this kind of support is often hard to find and can be misleading even when it is available. Indeed, as one British official suggested, comparative figures on direct versus indirect aid to be 'so meaningless as really not to be at all useful'. Another official noted that, 'there is no doubt that the financial position of the aircraft companies would look very different if they had to pay the full cost of work done at Farnborough and N.G.T.E. which is relevant to civil aircraft projects...[but] there is a very major defence interest here, and in this area it is sometimes very difficult to be sure where one stops and the other starts'. [13] Certainly, Airbus Industrie will cite the general research backing for its products which stems from publicly-financed R & D in each of its member states, in order to underline the strength of its products. But in general terms, the Europeans still claim that the difference in scale and degree of integration between their research effort and that of the United States continues to provide American manufacturers with a distinct advantage.

American success in achieving a dominant position in civil aerospace cannot be explained only by reference to the help Boeing, MDD and the rest have received from US defence spending. The past inadequacy of European civil aerospace has contributed to its poor commercial performance. Insufficient market research, indifferent product support, limited production, small, under-capitalised companies, and plain bad judgement, have all played their part. The central place of the dollar in determining prices has also meant that European firms have had to carry an exchange rate risk largely absent in the US, especially in the years when the dollar was overvalued against most European currencies. Higher European inflation has imposed an additional burden on European industry. These factors, combined with the fact that American investors appear to be more willing to take long-term risks on high technology, have undoubtedly put European civil aerospace on the defensive, and heightened its dependence on the state. [15]

Even if the role of US defence spending in explaining the success of American civil aerospace is overrated, it still provides a powerful political justification for European government support for civil programmes. Michel Lagorce, director of civil programmes in the French Ministry of Transport observed that 'in the United States, airliners are derived from military aircraft. Their development costs are just as much a claim on the state'. This view was shared by Martin Grüner, the German Federal aerospace coordinator, 'the US industry receives large-scale military orders from the American Government which provides it with quite different financing possibilities compared with European companies'. [16] Such statements serve to increase the politicisation of civil aerospace, and have fuelled the heated exchanges over subsidies with the Americans.

European support for civil aerospace has undeniably been both extensive and expensive. Since the Second World War, the British, French and later, the German Governments have provided launch capital for civil projects. All three of the 'Airbus states' use a mixture of repayable loans and credits to support the non-recurring costs of civil programmes, and production finance is often backed by government guarantees. Similarly, publicly owned firms may also be voted money or public loan capital which can be used to pay for civil work. The return on this investment has, to say the least, been poor. For example, between 1945 and 1974, British Governments spent £1504 million at 1974 prices on civil projects, and received less than £150 million in repayments. Together with the French, the British sustained an enormous loss on the Concorde. Between 1962 and 1977, every year the French Government spent, on average, £206 million on civil projects. Repayments averaged about £6 million a year over the same period. The Germans too, have experienced the financial effects of commercial failure in civil aerospace. Although there have been individual successes, with a few aircraft and engines earning money for both industry and the state, unprofitable projects have been all too frequent. Decisions to invest in 'uncommercial' projects have often been justified by reference to factors such as the balance of payments, technological spin-off, employment and sundry other political goals. This line of argument, and the fact that public money has enabled European companies to stay in civil aerospace when, it is claimed, American firms would have cut their losses, has led American companies to claim that they face 'unfair', subsidised competition.

The 'subsidised' Airbus

The subsidy issue has been one of the central features in the propaganda battle between Boeing and Airbus Industrie. During 1983, Boeing circulated calculations purporting to show the extent of the subsidies underpinning the Airbus programme. Boeing's data was derived from comparing the typical costs of developing and manufacturing American aircraft with the Airbus family, and showing various cash flow profiles over the initial years of development and sale. Boeing assumed that the average launch costs of a medium sized airliner were in the region of $1.5 billion and that 700 units would be delivered in the first 12 years, up to a total of just under 1000 over 25 years. If, as in the case of the L1011, a programme failed to achieve 300 sales in the first ten years, 'the probability of retrieving the investment becomes remote'. Boeing noted that Airbus sold only 57 aircraft in its first five years, and total sales for both the A300 and the A310 (in 1983) stood at a little over 350. Airbus direct costs, it suggested, were further increased by a longer development period than was commonly the case in the United States, and by lower productivity rates generally found in the European aircraft industry. Under such circumstances, even with the most favourable market conditions, it was unlikely that the A300/A310 series would repay the $10 to $12 billion which had been sunk in the programme. On the basis of past performance, Boeing claimed that 'A300 production would have been abandoned some time ago under US private industry criteria'. [17]

According to Boeing, the A320 will add a further $2 billion to Airbus Industrie's costs, and that it cannot be profitable at a price set by the market, even if it reaches its best target for sales in the first ten years of production. The total figure of subsidy, Boeing concluded, was worth roughly 25% on every Airbus sold. These calculations did not take account of other forms of assistance, such as export sales and the fact that the royalties levied by the governments on each Airbus, about $3 million on each aircraft sold, could be returned to the Airbus programme in the form of continued government support for new types and the development of derivatives. The intrinsic unprofitability of the A320 was, according to Boeing, underlined by the fact that Airbus has been unable to expand its membership, 'in spite of aggressive efforts directed towards Japan, Canada, Italy, Australia and others'. [18]

Airbus has received substantial sums from its governmental sponsors, in the form of direct aid or credits backed by government guarantees. *Flight International* estimated that the non-recurring costs of the A300 and A310, have exceeded £2000 million. The German Government, for example, has provided Deutsche Airbus with £767 million in long-term government loans, with interest payments deferred until 1994, and has guaranteed a similar sum to support loans raised from banks to help cover production costs. In the early 1980s, industrial repayments were also suspended as an emergency measure. German industry has had to find some of the money for Airbus development, but this has never amounted to more than 15% of the total German share. In the early stages of a project, it has usually been as low as 10%. [19]

A similar array of supports has been deployed by the French Government to finance the Airbus. Up to 1983 the French Government had advanced Aerospatiale £419 million for Airbus development. The French Government also provided capital and credit guarantees on a more general basis for Aerospatiale which it used to help defray production costs. The French Government covered all of Aerospatiale's commitment to the A300, but for subsequent variants, launch aid dropped in stages to 75% on the A310. The British Government, on the other hand, claims that it has not directly financed either the A300 or the A310. However, as a nationalised company, BAe received money voted under various Acts of Parliament to fund its share of the A310, as well as a one-off payment of £25 million towards Airbus 'work in progress'. [20] When BAe was privatised, these 'assets' and financial obligations were absorbed into BAe's own account. In 1983, BAe wrote off £100 million on its civil R & D account, much of which was attributable to the Airbus. BAe received £250 million in launch aid towards its estimated £437 million share of the A320. BAe will, however, have to find the balance from its own resources. The French Government is committed to provide 85% of Aerospatiale's £495 million share of the A320, and the German Government will advance Deutsche Airbus £395 million in aid and credits. The Spanish Government will provide all of CASA's share. [21] Overall, the programme is guaranteed against industrial default by the inter-governmental agreements signed at the start of a new project.

Naturally, Airbus Industrie, as well as its individual industrial members resent Boeing's accusations. BAe firmly rejects the view that any of its investment decisions are 'uncommercial'. The case it presented to the British Government for the A320 was based wholly on market projections and sound pricing assumptions. As Sir Austin Pearce argued, 'it was not a subsidy or a free gift. We will pay back the loan, with interest and profits, during the 1990s'. [22] For its part, the British Government argued that its investment in the A320 was indeed based on a fully commercial risk. The terms of the launch aid make it clear that BAe will have to meet stringent repayment schedules. In particular, one-fifth of the sum will have to be repaid by BAe over three years from 1990. In the past launch aid had been repayable only from the sales of individual aircraft or engines. Although BAe clearly believe that the Airbus will be profitable, outsiders have estimated that the rate of return is likely to be between 2% and 3% on BAe's investment, substantially lower than the average return in civil manufacturing generally, and about half that which BAe has predicted. [23]

Airbus Industrie believe that the thrust of Boeing's accusations is simply a response to unaccustomed competition. Royalty payments to the governments, for example, are not automatically transferred to the Airbus programme, although the French Government does admit that, as Aerospatiale's ultimate owner, re-investment is a legitimate policy option. [24] Equally, the assumption that American standards of efficiency are necessarily superior to those of Airbus are contested by Airbus Industrie. Airbus Industrie further deny Boeing's assumptions that the A300/310 will only sell 450 aircraft over 25 years. Currency movements constitute a major imponderable, and can make the difference between a profit or a loss on the programme. Indeed, currency fluctuation over long periods makes nonsense of comparative cash-flow predictions. In the specific case of the A320, Boeing has made no allowance in its forecasts for the fact that it benefits from an established infrastructure and the existence of a family relationship between it and earlier models. Moreover, Airbus Industrie argue that the A320 has achieved the kind of initial commercial breakthrough that will ensure viability. [25] Finally, the failure to expand Airbus membership is explained by the fact that many of the firms Airbus Industrie considered as prospective partners had long established links with American companies, and by other domestic political considerations.

Airbus Industrie claim that the A300/310 programme will break-even at 860 sales, a 'realistic' sales target over the life-time of the two aircraft. At that point, all government development loans will have been repaid. Already, Aerospatiale has repaid one-fifth of the money it has invested in the Airbus, and about $1 billion in total has been returned to the three governments. There remains the problem of paying off the accumulated debt incurred over the initial period of development when sales were very slow. Even generally favourable commentaries on Airbus financing have suggested that the member industries 'may never be able to liquidate the various combinations of capital infusions and loans received from their governments'. Clearly, the initial costs of establishing the A300, as well as the cost of building up the Airbus system with its associated infrastructure costs, has imposed a massive short to medium term burden on the programme. Significantly, even the French Government is looking for ways of financing civil aerospace on a more commercial basis. A $54 million loan to finance Aerospatiale's share of initial A320 development, put together by a consortium of French banks, will not be guaranteed by the state. [26] The launch of the A330 and A340 could add a further $2.5 billion to the total cost of Airbus programmes, and Airbus Industrie, as well as its individual partners, recognise that they cannot rely on their governments for the whole cost of launching these two projects.

In general terms, the French and their partners readily concede that without government help in the early 1970s, the project would have died. In this sense, they accept Boeing's broad case against the Airbus, or at least as far as the A300 and A310 are concerned. However, the French view the FFr 7 billion they have spent on Airbus as 'a ticket paid to challenge Boeing'. [27] Bob McKinlay, head of BAe's Filton Division, challenges America's right to regard civil aerospace as its sole preserve; 'in effect the Americans are just trying to say we in Europe can't fight our way back into the world aircraft market. But that's ludicrous. They are just trying to preserve their dominant position'. [28] As a German industrialist put it, Europe was buying more than just an aircraft, 'we are opening up the world commercial aircraft market for Europe, almost from scratch'. [29] From this perspective, the money invested in Airbus represents a strategic commitment to maintain a basic, and technologically important industry in Europe. British Governments might be less inclined to view Airbus in these terms, but the

French and Germans certainly see it as money well spent.

Detailing Airbus Industrie's financial status is notoriously difficult. The GIE system allows Airbus Industrie members to bury Airbus profits and losses in their own confidential accounts. However, as Airbus depends ultimately on public support, it is perhaps uniquely vulnerable to demands for a greater degree of public scrutiny. On the other hand, the accounts of US companies do not show the degree to which ailing projects, or commercially dubious sales practices are cross-subsidised from well established projects or from defence contracts. Airbus Industrie has argued that Boeing is able to exploit the monopoly held by the 747 to make attractive price offers for packages of Boeing aircraft, and to use its profits to cross-subsidise commercially weaker projects such as the 757. Similarly, when MDD signed a massive contract with American Airlines to supply MD80s, details remained strictly confidential, as have any losses incurred on other leasing arrangements. Boeing, as ever, was quick to provide a 'guestimate' of the cost of MDD's campaign which included selling the aircraft at between $2 million and $5 million below cost. [30]

It could be argued therefore, that there is precious little difference between 'normal' business practice, and the supposed subsidised practices of Airbus Industrie. However, as a publicly-supported programme, the Airbus perhaps should live by a different standard of accountability. But, as many industrialists continually argue, if you want the aircraft industry to be competitive, and have to operate successfully in a tough and ruthless environment, the professionals must be trusted to carry out a complex task with skill and due regard for financial probity. The conflict between public accountability and corporate autonomy is one of the classic problems of state assisted industry, and a satisfactory solution to the dilemma has yet to be found. [31] Airbus Industrie, however, will have to live with demands for greater openness in its financial affairs, and its members should be more sensitive to the public presentation of expenditure. The taxpayer must have confidence that his money is being used well and effectively. [32]

In the final analysis, civil aerospace in Europe is a public policy issue, whereas in the United States this is not the case—or at least not yet. As a result, any dispute between the United States and the Europeans will be underpinned by a fundamental difference of philosophy about the rightful place of the state in the promotion of civil industry and technology. John Newhouse has tried to put the conflict in perspective. 'On every side', he writes, 'there has been a good deal of complaining, much of it shrill....The Americans bitterly protest some of Europe's competitive practices, and the Europeans cite similar grievances of their own'. [33] Commercial aerospace is not the only technological issue dividing the United States and Europe, but it has been a particularly intense area of dispute. Aircraft have a high visibility, and the US industrial lobby has a powerful voice in Washington. The debate over direct and indirect subsidies in the development of civil aerospace increasingly resembles a proverbial dialogue of the deaf, with no obvious point of resolution. As one French official noted, the Americans are making a 'fundamental intellectual error in judging our actions and interests by their standards'. By the same token, European failure to appreciate the sensitivities of Americans on this matter, only invites suspicion and opposition.

Sales support policies

A second, and perhaps an even more controversial aspect of state support for the development of civil aerospace projects, is the help given to manufacturers by governments to sell their products. The sale of civil aircraft and engines has always been a hard and ruthless business, and as a number of scandals in the early 1970s revealed, outright bribery was not unknown. Although corrupt practices have been brought under control, or have at least become more discreet, there is still a whole range of inducements which can be deployed to secure a sale. All the major civil aerospace states have mechanisms and institutions to channel money, mainly in the form of export credit guarantees, which help overseas customers to purchase aircraft and engines from their national industries. These are especially important when the market is depressed, and when airlines have incurred high levels of debt. The last decade has also seen the advent of 'innovatory financing', involving leasing deals and 'buy-back' clauses in contracts. Overall, the market has seen a significant shift in the nature of civil aircraft sales financing, where the manufacturer assumes an equal, if not greater risk in financing airline purchases. In this context, the role of state support has become a major issue in the relations between American and European manufacturers. This is particularly the case where governments influence airline procurement, and are interested in obtaining wider commercial and political gains from a purchase. National leaders, well-connected politicians, diplomats and the military, have all played salesmen, or saleswomen, on behalf of their national industries.

The United States, Britain, France, West Germany and Japan, all have agencies designed to support the export of manufactured goods. Britain and the United States in particular, have a long tradition of providing soft loans and export credit guarantees to private institutions for this purpose. The origins of the British Export Credit Guarantee Department (ECGD) can be traced back to the Overseas Trade Act of 1920. Similarly, the United States Export–Import Bank (Exim) was founded in 1934 as part of the New Deal. [34] COFACE and Hermes, the French and German equivalents, are post-war creations, but they too reflect a long-standing philosophy of supporting export sales. Civil aerospace has traditionally attracted a considerable proportion of the export credits given by these institutions. British and American commercial aircraft and engines accounted for between a quarter and one-third of ECGD and Exim funding during the 1970s. From 1957 to 1975, Exim authorised $5.4 billion in direct credits to support $13.6 billion in export sales, and in the fiscal year 1979–80, the ECGD provided an estimated £537 million in support of British civil exports. [35]

While the details of national mechanisms and relations between governments, credit agencies and private lenders, vary from state to state, the function of an export credit system is much the same in all cases. [36] Customers are helped to defray the high initial cost and long amortisation period of investment in civil aerospace purchases. Loans and credits are usually offered at terms substantially better, both in respect of interest and repayment time, than could be obtained, if at all, from the commercial market. For example, in 1978, Laker borrowed $131 million over ten years at 10.2% when the commercial rate was 13%. [37] As one leading authority on export credits observes, 'the basic ingredient of official export finance is an export credit that embodies an element of subsidy in the form of a lower rate of interest, or a longer repayment period than would be available on a commercial loan'. [38]

Governments are also prepared to take on customers with a high risk of default, whether impecunious airlines, or, more significantly, unstable and economically vulnerable countries. Both the general principle, and specific cases of export support have attracted considerable political controversy. On the one hand, critics of the system argue that governments should not distort the international market by subsidising the sale of aircraft and engines, particularly when so much aid goes to a limited number of beneficiaries. On the other, manufacturers have bemoaned the lack of support given by their national credit agencies in the face of more extensive aid available to their competitors. In recent years, a dispute has emerged between the United States and the Europeans about the terms on which such aid is offered and the extent to which the Europeans are offering 'predatory' finance to buy into the US, or American dominated markets.

The European and American dispute

In general, European manufacturers suggest that American complaints about export credit policies are a reaction of someone tasting his own medicine for the first time. The advantages which American manufacturers were said historically to possess were outlined in a British Parliamentary Report of 1964. British firms then claimed that the Americans had a 'distinct advantage' because of more extensive and flexible export credit facilities they received from Exim. Although the Committee was 'unable to discover any specific cases where orders have been lost as a direct result of the unavailability of attractive credit facilities', it did concede that marginal advantages in terms of interests rates might be decisive in a close competition. Moreover, according to British industrialists, the Americans were generally more adept at harnessing their diplomatic and political agents in support of US aircraft; 'all their ambassadors, their commercial counsellors, their Air Force officers, everyone talks American, and they back up their goods'. Equally, American manufacturers were helped by 'a generally superior financial system, deeper capital markets offering longer maturities and often lower interest rates'. Exim could then help additionally by financing cases where commercial finance was constrained. [39]

The current dispute was triggered by a combination of the increasingly tough market conditions of the late 1970s, and by the onset of more aggressive European sales offensives. One of the opening shots was Airbus Industrie's bold strategy to win its first American order from Eastern in 1978. Airbus Industrie, and certainly the French Government believed that Airbus had lost an earlier American customer as a result of 'buy American' pressure; all concerned were determined this time to make Eastern an offer it would find too attractive to refuse. In effect, Airbus offered Eastern four A300s on a free trial for six months, running the risk that the airline would return the aircraft with very damaging consequences for Airbus' reputation. To reduce that risk, Airbus's Roger Béteille formulated 'an operating cost guarantee' which compensated Eastern for the difference in operating costs between the Airbus, with 240 seats and Eastern's specified 170 seat requirement. Eastern were also granted favourable export credit terms. [40]

The concern felt by American manufacturers about the Eastern order was aggravated by another case involving Panam, Rolls Royce and the British Government. Panam had ordered 12 Lockheed L1011s, and although Rolls' monopoly on the L1011 had not been challenged, Rolls was worried that Panam might select an American engine for its L1011s. The Panam Board had already accepted engineering opinion that another engine would be technically and economically unjustifiable. However Panam was in desperate financial trouble, and wanted the best possible deal. Panam discussed the matter with Rolls, implying that GE had proposed a package which included the aircraft as well as the engine. Rolls felt it had to respond and appealed to the British Government for help. With the personal intervention of the Prime Minister James Callaghan, the ECGD was authorised to extend a £250 million credit covering both the engines and the aircraft. The reaction from both Rolls' American competitors and the US Government was predictable. The British had broken all the rules, albeit unwritten ones, which were supposed to govern export credits. In the view of one P & W official, 'how can we compete with the British Government?' Sir Kenneth Keith, Rolls' Chairman, said that there was nothing in the package that 'contravenes the fair trade rules'. Nevertheless, the Panam deal only reinforced a growing belief in the US that Europeans would stop at nothing to win sales. Even when the airline's dubious role in the affair became clearer, it made little difference to the poor impression made in the United States by European sales tactics. [41]

The main cause of American concern was the growing depth of Airbus Industrie's market penetration. In 1976, Airbus Industrie had some 2% of the market for medium-range, wide-bodied airliners; by 1981, this had risen to 55%. Although some way short of Airbus Industrie's objective of 30% of all airliner sales, it was increasingly evident that this target was not just wishful thinking. The threat was made worse by the precipitate slump in the market which occurred between 1979 and 1983. Falling traffic, suicidal price competition, and airline mis-management all took their toll of airline liquidity. The accumulated debts of most airlines, with extremely high servicing charges, meant that the pressure was on to defer, or even to cancel orders of new equipment. Indeed, one consequence of the recession, with a number of large carriers going to the wall, was a glut of 'nearly new' second-hand aircraft. A low hours DC10-30 could be had for $25 million, against a new Boeing 767 priced at $45 million. As one airliner salesman put it, 'some airlines are telling us, "to hell with long range plans"' and were taking up older, but cheaper equipment. [42] In these circumstances, manufacturers were desperate to win even small orders to maintain production levels and their sales momentum. One effect of the crisis was the appearance of 'innovatory' marketing strategies, such as leasing and 'buy back' arrangements. These will be considered in more detail below. Another was the increasing importance of governmental involvement in the sale of civil aircraft and engines. Where the Airbus–Eastern deal and the Panam engine affair had been straws in the wind, the deepening recession was to raise the proverbial hurricane.

The focus of American complaints was the success Airbus had achieved amongst the so called 'silk route' customers of the Middle and Far East, traditionally a strong Boeing market. Airlines in this area were significant for a number of reasons. First, the local pooling of facilities tended to benefit an initial breakthrough; a sale to one, regionally influential airline tended to have a knock-on effect on others in the pool. Secondly, most airlines in the area were weathering the recession rather better than most, and had hard cash to spend on new equipment. Finally, many of the airlines were subject to government control, and vulnerable to government to government negotiations involving wider political issues. By the early 1980s, Airbus Industrie had won a substantial share of this vital market. Undoubtedly, its products were good; in particular, many airlines discovered a 'sleeping' market in air freight thanks to Airbus's superior cargo carrying capacity. However, American firms began to complain about Airbus Industrie's 'uncommercial' tactics and the political inducements offered by its government sponsors, especially the French, to buy the Airbus. For example, Boeing claimed that an important order from Kuwait had been cancelled in favour of Airbus as a direct result of an inter-governmental deal involving the promise of French investment in the Kuwaiti petrochemical industry, and extremely favourable financial terms granted by the French Government. The 'Airbus states' also appeared to be prepared to accept risks which the American Government refused to underwrite. [43]

Although some cases of European 'sharp practice', were in fact, a commercial backlash against American foreign policy in the region, governmental involvement in the sale of aircraft and engines rapidly became a major issue in the United States. [44] In 1978, in evidence to a House of Representatives sub-committee, J. B. L. Pierce, Boeing's Treasurer, cited Airbus Industrie and the support it received from governments, as a clear example of a 'predatory export financing scheme'. Airbus tactics, he said, showed just how far 'European manufacturers and their governments are willing to go to penetrate new aircraft markets'. Boeing could compete on technical merit with the best in the world, but not against 'the national treasuries of France and Germany'. Pierce called for, if not overt retaliation from the United States, at least similar help from a more aggressive American export credit policy. [45] The problems over civil aerospace were increasingly bound up in a wider set of trade related issues which, by the late 1970s, were beginning to affect Euro-American relations. However, given the history of American domination of the civil aerospace market and the political clout of the US aircraft industry, with its state-wide network of subcontractors, its complaints had a high political profile. Although the debate was complicated by mutual participation in each other's civil programmes, the issue polarised around the fight between Airbus and Boeing, and on the issue of government support for their respective sales campaigns.

The GATT agreement on civil aircraft

The most immediate focus for discussions about trade in civil aerospace products was the GATT. At the insistence of the US, civil aerospace was included towards the end of the protracted negotiations to secure the 'Tokyo Round'. The 'Aircraft Agreement', signed in 1979, aimed to eliminate all customs duties hitherto levied on civil aircraft, parts and repairs. It also contained provisions to prevent the use of technical standards to give domestic products unfair advantages; to ensure that domestic customers only bought aircraft on commercial and technological grounds; and prevented the imposition of import quotas. The Agreement additionally sought to limit the role of 'non-tariff' barriers to free trade in civil aerospace. [46] The 'Tokyo Round' revealed quite clearly the different perspectives of the two main protagonists. The Americans targeted their efforts on the abolition of government support for development and export subsidies. The EC Commission, acting on behalf of the European countries, aimed more specifically at the 'Buy American' legislation which obstructed European penetration of the US market. [47]

Export credit policies, as well as development aid, clearly came within an American definition of non-tariff restrictions on trade, but the 'Tokyo Round' fell somewhat short of American hopes. The Aircraft Agreement recognised that government financial support was widespread in the aerospace industry, and that many states, especially the Europeans, viewed this activity as an essential part of their industrial and economic policies. The Aircraft Agreement called upon all parties to ensure a 'reasonable expectation of recoupment of all costs, including identifiable pro-rated costs of military research and development (programmes) that are subsequently applied to civil aircraft'. [48] How this was to be done, and how one might define 'reasonably' was left vague; as was a similar requirement to end 'unreasonable pressure' on national airline procurement. No agreement was reached on the scope of government export credit support. [49]

The problem for the GATT negotiators was that the civil aircraft industry was so bound up with a wide range of governmental interests, not the least of these being the health of important defence contractors. Indeed, Article 6.1 of the Aircraft Agreement stated that it had to 'take account of the special factors which apply in the aircraft sector, in particular the widespread governmental support in that area, their international economic interests, and the desire of producers of all signatories to participate in the expansion of the world civil aircraft market'. [50] The dilemma was that the stimulus for real competition in the civil aerospace market lay in 'the very support given by European governments to their aerospace industry which has permitted the latter to regain a place on the market...notably through the development of the Air-bus'. [51] In short the GATT Aircraft Agreement largely reflected the European view on non-tariff barriers. A Committee on Trade in Civil Aircraft was formed to review the position on an annual basis, but the GATT left the establishment of an effective international regime to bilateral negotiations between the aerospace producers. [52]

This hardly satisfied American industrial interests. The US industry representative to the GATT talks, George C. Prill, late of Lockheed, accepted that the details of an agreement on non-tariff barriers would take some time to work out, but he warned that 'if no compromise can be reached, President Carter could withdraw the US from the agreement'. He hoped that this would not happen. The US, after all, was dealing with 'friends not enemies'; but the hint of a possible trade war was clear. [53] With the United States facing a growing problem with a balance of payments deficit, and increasing European penetration of civil aerospace markets, President Carter came under pressure to promote American exports and to curb the perceived excesses of the Europeans.

In the short term, the Americans found it impossible to persuade the Europeans to adopt an acceptable regime covering civil aerospace exports and development subsidies. Carter therefore, decided to take the offensive, and in the dying months of his Administration, proposed measures designed to enable the United States to compete more aggressively in export markets. Exim was to play a major role in this policy, but how it was to do so without relaxing its long-established practice of remaining financially solvent while increasing substantially its subsidies to United States exports, was not resolved. Indeed, as Joan Pearce put it, a policy to integrate the public and private sectors into a 'concerted endeavour to increase exports was something that ran counter to the private enterprise philosophy of the United States, and for which it was ill-equipped'. [54]

Negotiating the Consensus

The arrival of Ronald Reagan and 'Reaganomics', intensified the contradictions in American official attitudes towards industries like civil aerospace. Its importance as a national asset was recognised, but as it already benefited from forms of public assistance which were anathema to the hard-line free marketeers of the Reagan Administration, requests for more help were not sympathetically received. Indeed, the new Administration was set on reducing, not increasing aid to the civil aircraft industry. For example, in 1982, Reagan authorised the repeal of the 'safe harbor' tax credit system which enabled unprofitable airlines to sell unused long-term tax reliefs on the purchase of capital equipment to profitable carriers, which they could then use to finance their own aircraft purchases. [55] The proposed abolition of 'safe harbor' triggered a frenzied campaign by both the airlines and the aircraft manufacturers to get the decision rescinded, or at least to have it phased out over a number of years. In the event, the government agreed to change the tax credit system in stages, with generous terms for airlines which had already placed orders for new aircraft on the basis of 'safe harbor', or whose procurement plans were well advanced. The threatened cancellation of orders did not occur, but the repeal of 'safe harbor' indicated that the United States aircraft industry would have a hard fight to convince the new Administration that it needed help to beat European competition. [56]

Indeed, rather than responding to the 'European threat' with a more aggressive export financing policy, the US Government intended to reduce Exim's funding by 12%, and the bank was instructed to remain profitable. To ensure this, Exim would have to increase its interest charges in line with commercial rates, already beginning to rise dramatically under the general impact of tight monetarist policies. Worse still, the Reagan Administration had ideological reservations about the role of Exim in the economy. The belief was that its lendings distorted the flow of domestic capital, 'crowding out' other borrowers who would use the money more productively. The fact that the aerospace industry, especially Boeing, taking on average 40% of Exim credits, dominated the system, was further reason to resist pressure for an increase. Eximbank's new President, W. H. Draper, was also an opponent of 'subsidisation'. He felt that the bank should remain self-sufficient and 'defensive' in terms of countering foreign export credits. [57] The aerospace industry's reaction was predictable; the Reagan Administration did not understand the competitive context of the civil aerospace business, and had failed to appreciate the extent of foreign subsidies.

Even if the United States government would not directly counter European subsidies, it could try to persuade the Europeans to follow commercial practice in respect of export credits. This, of course, was the crux of the matter. The Europeans had to be assured that the Americans would not unduly benefit from the greater scale and flexibility of the United States capital market. [58] Moreover, they would have to make most of the concessions on interest rates and loan periods. Some Europeans, notably the French, felt that as the Americans had begun the 'credit war' with Carter's initiative, they would have to face the consequences. [59] However, French aggression was balanced by British and German fears that conflict over export credits for civil aerospace could damage general economic and political relations with the United States. In August 1981, a truce was negotiated between the United States and the 'Airbus states'. The so called 'Commonline agreement' fixed export credits for civil aircraft at 12% over 12 years, with direct credit support ranging between 62.5% and 42.5% depending on the repayment schedules of private and official financing. [60] The Commonline entailed some limited concessions from the Europeans, acting in response to American threats to impose general import duties on countries which were regarded as excessive subsidisers of exports. The French again wanted to fight for better terms, but the Germans, supported by the British were prepared to accept a temporary agreement pending more detailed negotiations on aircraft export financing. The agreement was to run until October 1982. [61]

Despite a general welcome for the Commonline agreement, American manufacturers still believed that the United States Government could do more to stem 'predatory financing'. The Europeans were believed to have won on the Commonline. The French were still believed to be undermining the spirit, if not the letter of the agreement, by offering other inducements and incentives to win Airbus sales. Despite some threats from the Administration to impose import duties, American industry felt that more should be done to curb European malpractices. At the very least, it could make it clear to the Europeans that, *in lieu of a* firm undertaking on export credits, the United States would retaliate with its superior resources to 'deep pocket' the opposition into agreement. In short, Exim should be encouraged to play the Europeans at their own game by offering more competitive and agressive financial packages. Neither the Administration nor the Exim bank was impressed. A few concessions were made to the manufacturers of 'moderate cost' equipment, and Exim officials said they would respond 'when it is truly needed to meet the foreign competition'. However, the Bank did not see its role as providing a stimulus for US production when demand was sluggish. The Bank's President, William H. Draper was even more sanguine; 'we're honestly getting to the point where all countries are recognising the folly and the lack of effect of these subsidy policies'. [62]

If the aerospace industry found it hard to impress the Federal Government, Congress was more receptive. The question of free trade and protectionism was undoubtedly a live issue in both the Senate and the House of Representatives. Pressure was growing across a range of sectors to protect American industry, and Boeing for one was not slow to alert a wider political constituency to the dangers of lost markets. The company orchestrated its national network of subcontractors to advertise the threat to American jobs and sales presented by subsidised foreign products. [63] Moreover, the fact that the Exim charter was up for renewal in 1983 provided a specific focus for Congressional pressure. Although members of both Houses, either in sympathy with 'Reagonomics', opposed to protectionism, or simply against aid to giant corporations, to some extent counterbalanced the industrial lobby, it could count on formidable support from politicians with aerospace interests. As a result, throughout 1982, various proposals were tabled in the Senate and the House to provide Exim with a 'war chest' with which to meet foreign competition. [64]

The noises from Congress, as well as the continuing complaints from the industrial lobby began to have some effect on the Administration. More important perhaps, there was little sign that the Europeans were prepared to make any more concessions on export credits. During September and October 1982, officials from the United States, France, Britain and Germany held a series of tense and highly confidential meetings. The Americans wanted to move as quickly as possible to a fully commercial regime, with interest payments reflecting commercial rates. The Europeans, however, were determined to stay within existing guidelines. As a British sokesman put it, although the Europeans were willing to consider adjustments, certainly in relation to modest changes in the length of loans, 'they do not want things to go too far and get out of hand'. [65]

In the event, the parties could only agree to extend the Commonline pending further negotiations. Public statements were guarded, but it appeared that the Europeans had successfully resisted American demands for substantial reform. The United States Treasury said that the negotiations were a 'watershed' and that it was hopeful of reaching eventual agreement. However, Draper claimed that the Europeans had not accepted 'any change at all' and that the failure to move rapidly towards commercial rates was 'disappointing'. Robert Lee of the United States Commerce Department was equally pessimistic; 'nothing was concretely resolved'. In his view the only reason why the United States should not unilaterally abrogate the Commonline was that 'some agreement is better than none'. [66] In the absence of progress multilaterally, the American Government began at last to hint at stronger unilateral measures. In January 1983, it was announced that Exim loan authority would be pegged directly to the size and extent of foreign subsidies. Although there would be no immediate increase in the authority sought for Fiscal 1983, the United States Government served notice that it might seek additional funds if this was needed to counter 'inappropriate' foreign government export financing procedures.

The aerospace lobby in Congress wanted even more. As the Exim Charter wended its way through Congressional renewal, various bills were presented to strengthen the Bank's mandate to match foreign subsidies. Some Senators wanted to clarify the priorities which Exim should follow. It was felt that the Bank should pay less attention to remaining 'commercial', and more to the active promotion of United States exports. As one Senate Committee economist put it, the thrust of the proposed legislation was 'a mandate to neutralise foreign export credits, making them a non-factor in United States export competition'. Indeed, it was to be expected that these would be 'the marching orders for the Bank to go out there and aggressively compete'. Others were prepared to go further still: one Congressional proposal aimed to link Exim and AID activity, including a $1 billion 'Competitive Tied Aid Fund' to be administered by Exim. For its part, the Senate was prepared to consider providing for direct sanctions against 'predatory financing', with United States companies able to petition for bans on the import of subsidised products. One Senator suggested that the regulatory work of the Federal Aviation Authority should also be linked to the promotion of United States manufacturing interests. This was too much even for supporters of the United States aircraft industry, and was roundly condemned by *Aviation Week*; 'involving the FAA in trade battles...is an invitation to bigger troubles in the export field'. [67] It was, however, an indication of growing Congressional dissatisfaction with the Administration's policy.

In November 1983, Congress voted to renew the Exim Bank Charter. The new Charter bolstered the bank's mandate to be fully competitive with other national export support policies, and included the tied aid provision. The new Charter also clarified the Bank's priorities between solvency and competitiveness; although the average cost of money was to be one factor in the bank's operation, this should not 'impair the bank's primary function of expanding United States exports through fully competitive financing'. Where there was a clear challenge to commercial rates from a foreign agency, Exim was authorised to 'neutralise the effect of such foreign credit on international sales competition'. [68]

Internationally, the Commonline was periodically renewed, but progress towards achieving a more permanent agreement was glacial. However, the discussions were being affected by a changing commercial environment. There was a growing feeling that a more sensible approach to export credits would be to everybody's benefit. In Europe, the threat of an all-out trade war in civil aerospace was an effective deterrent to unrestricted export credit policies. Although the Europeans had presented a united face to the Americans, differences between the French and their partners were increasingly apparent. The British and the Germans were not only more alert to the dangers of pushing the Americans too far, but were also more ideologically and commercially sensitive about the cost of export subsidies. Although the French were less affected by these concerns, the recovery in the civil aerospace market led to increased confidence in the prospects for an agreement which would resolve French fears of the Americans gaining too much advantage from a movement towards a more commercial regime.

In July 1985, an agreement was signed which set the length of repayment at 12 years, and introduced a flexible system of interest payments. The fixed rate of 12% specified by the Commonline was replaced by one which varied automatically on changes in the yield of official bonds. In the US, the minimum interest rate was related to US Treasury stock; in Europe the level was based on a Sterling, Deutschmark and French Franc 'currency cocktail'. Effective rates would vary according to whether the period of loan was 10 or 12 years in duration. The governments also agreed to prohibit the mixing of export credits with other forms of aid credit. Significantly, the Europeans accepted American proposals to ban official support for sales in their domestic markets. There were no changes in either the minimum cash payment required for an export credit (15%), or the maximum price to be supported by credit (62.5%). The agreement covered all aircraft over 70 seats. [69]

Reaction to the agreement was muted. In the interim, the Commonline had held for some important deals. In particular, Panam's Airbus contract, although containing a number of obscure features, had been carefully scrutinised by US Department of Commerce officials, and no irregularities could be proven. There was still an undercurrent of unease in US industry circles about the future role of Exim, especially about proposals to eliminate its direct loan authority in favour of a complex system linking its guarantees to private institutional loans. [70] However, improving market conditions had moderated some of the fear of European competition, and most companies were prepared to get on with the business of selling aircraft. The broader question of European governmental support for development was unaffected by the international agreement on export credits, and US concern about losing long term competitive advantage would nevertheless remain a potential source of conflict between the United States and Europe.

The changing pattern of civil aerospace sales financing

The international politics of export credits and development subsidies has to be viewed against a general change in the way civil aerospace products began to be sold. The recession of the early 1980s spawned a whole range of 'innovatory' financial strategies to sell airliners and engines. The traditional way of buying airliners was for the customer to put down a cash deposit and to pay several pre-delivery instalments, the balance being paid on delivery. Manufacturers have now begun to assume a greater part of the risk in financing sales. Similarly, with airlines often short of cash, or carrying huge debts, in a context of technological uncertainty, they have sought to delay firm orders in favour of leasing aircraft. In some cases, barter deals involving oil and participation in airline equity have been used, in motor industry parlance, to 'shift metal', or at least to keep in touch with a prospective customer. Although some of these measures were regarded as short-term expedients, many have become a permanent feature of the civil aerospace business. As one Boeing executive put it, 'financing is now an integral part of the marketing process'. [71]

The most dramatic examples of 'innovatory financing' were MDD's leasing agreements with TWA and later, American Airlines. MDD were leasing Super 80s at between 25% and 40% of their price over five years. These terms were generally regarded as a give-away by MDD, but the rationale was to maintain production momentum and MDD's presence in the civil market. Although MDD hotly rejected accusations that it was buying a market, the company set aside $700 million to support its campaign. [72] MDD's strategy proved remarkably successful. Its original leasing deal with American was followed by a firm order worth up to $1250 million, potentially the largest airliner order in history. Neither MDD nor American revealed details of the contract, but American probably received up to $6 million in price concessions on the first 20 MD80s. MDD also accepted reduced pre-delivery payments, modest cancellation penalties, and improved the terms of American's initial leasing arrangements. [73] At first, Boeing refused to follow MDD's lead. However, in January 1983, it too concluded a complex deal with Delta to lease 737-300s. Airbus Industrie also offered to lease on BA's behalf Boeing aircraft if it ordered the A320. In the event, BA concluded its own leasing arrangement with Boeing in order to keep its options open on new technology. [74]

The system of manufacturer financing has become more formalised with the creation of company credit agencies and by associating with established finance houses. MDD's own agency was set up as early as 1970, and in 1985, it had $1 billion in outstanding loan and lease financing. In 1984, MDD also signed an agreement with Guiness Peat to lease its aircraft. Boeing now has a similar holding worth $600 million, and has an arrangement with the Miami based International Lease Finance Corporation. Airbus Industrie has also entered the arena with a sale of A320s to the Australian airline Ansett, which are to be leased to other airlines on delivery. [75] In order to win new orders, all three manufacturers are equally willing to buy back not only their own products, but those of their competitors, and have specialised divisions to sell and to support these second-hand aircraft. [76]

In terms of the battle between Airbus Industrie and the Americans, these trends only serve to complicate the whole issue of subsidised sales. The difference between MDD's tactics and those of the supposedly 'predatory Europeans' may seem to be only a matter of semantics. Nevertheless, Airbus's most dramatic penetration of the US market, its $2 billion order from Panam in September 1984, had considerable impact on American opinion. The Panam contract, albeit hedged with every manner of concessions to Panam's parlous financial state, was rightly regarded as a major coup by the European consortium. As the unofficial 'flag carrier' for the United States, Panam had never before bought an aircraft from an non-American manufacturer, and the total order was one of the largest ever trade deals between the US and Europe. [77] The initial order for 12 A310s and 16 A320s was alone worth £760 million, but if Panam exercised options on a further 13 A310s and 34 A320s, the total would be at least £2 billion. Airbus Industrie, conscious of the possible political repercussions in the US, was quick to stress the American content of each Airbus. [78]

American reaction was mixed. Wall Street felt that Boeing had been right to stay out of a price competition with Airbus, and its shares rose to a 1984 peak. Boeing's official reaction was that they had made a commercially viable offer to Panam, with aircraft which it believed met the airline's needs. 'We have to make a profit to survive', said one spokesman, implying that Airbus Industrie's offer was far from being a commercial proposition. American stock analysts were more forthright: the deal was based on substantial price concessions and the fact that 'Airbus wanted the deal so badly that it was willing to make an offer that Boeing could not match as a publicly owned company'. However, as hard as Boeing and US Commerce Department officials tried, they could find no conclusive evidence that the deal had breached the Commonline. [79] According to Bernard Lathière, 'not one centime, not one penny, not one Deutschmark of European tax-payers' money is involved in the deal'. Pierre Paillert, the man who had led the Airbus negotiating team, reinforced this point; 'this is a very standard deal. We haven't accepted coffee beans, cocoa or bananas in return for the planes, and we're not taking back any old Panam aircraft'. [80] No export credit was involved in the deal, and International Aero-engines, whose V2500 was selected to power the A320s, would also help to finance the order.

Outside observers felt that despite Airbus denials, the order had been secured by a sophisticated financial package which must have taken account of Panam's acute financial position. Although figures for Panam's most recent financial performance had been encouraging, it still had the lowest US credit rating for a viable business. The prevailing strength of the dollar would have helped Airbus, but Lathière refused to be drawn on exactly how the purchases were to be financed, especially as the order stretched over nearly a decade; 'this will depend on the money markets and other considerations at the time'. It was Panam's responsibility, and the money could be raised on American, European or Japanese capital markets. There was a possibility that Airbus would provide some form of partial guarantee, albeit at a cost to Panam—a procedure not unknown to either MDD or Boeing. This might have included a mixture of operating and financial leasing, where Airbus retained a share in the equity of the aircraft used by Panam. Panam may also have asked for guarantees of the second hand value of the aircraft, perhaps with an insurance policy taken out at a premium by the vendor. As a London banker put it, 'whatever the terms of the order, I am sure they will turn out to be extremely favourable. There must have been some incredible giveaways'. [81]

Final confirmation of the Panam order took some time, and it was not until May 1985 that Panam and Airbus Industrie signed a definitive contract. [82] Nevertheless, Airbus Industrie was jubilant about its breakthrough. It was presented as a clear vindication of the faith placed in the whole programme, and especially in the A320. It was especially satisfying to beat Boeing on its home ground. Reaction in France was particularly ecstatic; Dominique Baudis, the European Parliament Member for the Toulouse area, said that the order was 'a grand victory for Toulouse, for France and for Europe'. [83] In terms of the wider politics of international civil aerospace, the deal marked the final coming of age of the Airbus consortium and reinforced its presence as a powerful competitor to American industry.

Subsidies and civil aerospace, a continuing issue

In general, the return of a more buoyant market took some of the heat out of the dispute between Europe and the United States over government support for development and sales. However, such were the stakes involved in the battle between Airbus Industrie's A320 and the American derivatives, that any major order could be surrounded in controversy. This proved the case when Air India decided to cancel a letter of intent with Boeing to buy 757s in favour of an improved offer from Airbus Industrie. [84] Moreover, as civil aerospace moves towards another phase of technological and commercial development, the debate over subsidised civil aerospace is likely to return with an increased intensity.

As so often in the history of commercial aviation, events are being forced by developments in propulsion technology, in this case, the potential of the propfan/UDF engine. [85] There is considerable debate about exactly when a UDF airliner will be in service, and technical predictions are affected by an intense commercial propaganda war involving both the airframe and engine manufacturers. The dilemma facing all the participants is again the classic problem of timing the introduction of new technology to catch the surge of orders for 130 to 160 seat airliners predicted for the next ten years. The stakes are very high indeed, and a correct choice could very well guarantee commercial viability into the next century. The position is 'all the trickier' because each manufacturer 'is the captive of its past'. [86] The stance adopted by each company regarding the UDF is being largely shaped by existing commitments, and its current range of products. Airbus Industrie, having recently launched the A320, claim that the UDF will not be in full commercial service until early in the 21st century. In the meantime, the A320, with its substantial savings over current airliners is the best buy. Boeing and MDD, on the other hand, offering cheaper derivative aircraft based on designs conceived in the late 1960s, argue that the UDF will be flying much earlier than that, and airlines should wait for the huge increase in efficiency promised by the UDF airliner in the 1990s. Both Boeing, with the 7J7, and MDD, with the MD91X, have prop-fan aircraft under way. MDD hope to fly a demonstrator airliner by the summer of 1987, and the two companies have also begun to line up partners to share the costs of development. [87]

The engine companies are similarly divided. GE, with its French partner Snecma, have already started advanced development of a UDF engine. P & W agree that the UDF will be available in the 1990s, and also has a UDF research programme in hand. Rolls Royce is more cautious about the UDF. This is partly due to the fact that the Americans have invested more in UDF research, money which Rolls cannot match, but Rolls also has fundamental doubts about the UDF concept. The problem for Rolls is that the others are betting each way on the UDF. Snecma and GE could fall back on developments of the CFM56, and P & W has a share of the V2500. Rolls, however, can only afford a limited research programme without further aid from the British Government. Rolls, of course, could, at some stage, try to join either P & W or GE in a collaborative programme, but the further they get ahead of Rolls in UDF technology, the harder it will be for the British to join as anything like equal partners. [88] A lot, therefore, depends on the progress made by the various UDF research programmes. The health of companies on both sides of the Atlantic will be riding on events over the next five years. David Learmont of *Flight International* summed this up very nicely; 'in the battle for sales between the Europeans and Americans in the 1990s and beyond, it seems as if victory or defeat rests almost entirely on that single revolutionary step in the propulsion field. Prop-fan is a high risk, high reward item on which manufacturers are gambling the futures of their mid-range civil aircraft programmes'. [89]

Boeing has used the prospect of the UDF airliner to contrast further its private investment in the 7J7, which could amount to something in the region of $4 billion, with the subsidised Airbus. Pointing to Airbus Industrie's appeal for government aid to finance the A330/340 programme, Boeing again called for US Government action to counter 'state-funded capitalism'. In September 1985, the US Government raised the issue of development subsidies at a regular meeting of the OECD, but its call for an end to all aircraft subsidies was rejected by the EC representatives. In a collective letter, the Airbus governments denied all charges, arguing that Airbus financing was 'defensible' according to GATT rules. They, in turn, proposed informal talks over all forms of aircraft financing with the intention of pressing Boeing and MDD into revealing details of their financial practices, and to publicise the 'unfair support' US manufacturers received from defence and space programmes. However, the Americans made it clear that they intended to keep the issue high on the agenda of future trade talks with the Europeans, and would seek a more effective implementation, 'both in letter and in spirit', of the GATT Aircraft Agreement. [90]

In searching for an adequate response to the European 'challenge' the Americans were, however, still trapped by their own prejudices against direct state intervention to support the civil aerospace industry. For example, Boeing wanted action to prevent the erosion of the 'free market' in civil aerospace, but, as one executive put it, 'I don't support any form of subsidy for us.' With its considerable export market, Boeing also opposed protectionism and possible retaliation. [91] Equally, despite noises about establishing the export 'war chest', the Administration's actions continued to belie its words. Further changes to Exim's status, designed to 'privatise' its loan system, were criticised by both the aerospace industry and by American banks, which were to be expected to carry more of the risk entailed by financing export credit. One analyst of the US Exim Bank went so far as to suggest that the whole exercise of 'war chesting' on the part of Congress or the President was always likely to be more demonstrative than real. Without a consensus on the matter within the Administration, and a willingness on the part of Congress to finance a substantial increase in export support, the United States would never be able to mount an effective counter strategy for the Exim Bank. [92]

Nevertheless, the propaganda war has continued. Tom Bacher, for example, asserted that there was no economic reason why Europe had to have its own aircraft industry; 'every country does not have to build every product it consumes. You build good train systems and things like that in Europe, and we do not'. [93] Such sentiments cut little ice in Toulouse; according to Jean Pierson, even if Airbus Industrie does not win every sales battle with Boeing, 'it is still good to see some of Boeing's blood spilled on the carpet'. [94] The strategic view of aerospace still holds for Airbus Industrie. The expansion of its family of aircraft is seen as a basic requirement for European civil aerospace. Although launch finance for Airbus' new aircraft may not be as easy to find as in the past, the strength of European political commitment to the programme also remains high. In the struggle for commercial supremacy, perhaps even for survival, in the civil aerospace industry, the state will clearly continue to play an important role. The question for the future is whether the US will finally resolve its internal agonies over intervention and strike out positively to take on the strategic outlook of the European civil aircraft industry. There are indications, for example the interest in developing a hypersonic transport for both military and commercial use, that the United States is beginning to put together a technological counteroffensive. This, combined with the private and public investment in prop-fan engines, may yet prove to be powerful reposte to the Europeans.

Notes

1. Bluestone, op.cit., p.170.
2. Lockheed's problems went deeper than just problems with the L1011, and even this was hardly helped by the bankruptcy of its engine supplier, Rolls Royce.
3. E. Fitzgerald, *High Priests of Waste*, New York, 1972, p.315.
4. M. Edmonds, 'Market Ideology and Corporate Power in the United States', in K. Dyson and S. Wilkes (eds.), *Industrial Crisis*, Oxford, 1983, pp.81–7.
5. Newhouse, p.118.
6. See Hayward, pp.99–104. See also Lord McFadzean's evidence to HoC 389, 1981–82, Q345.
7. *AWST*, 21 May 1984, p.36; *Interavia*, September 1984, p.916.
8. G. Eades, 'Government Support for Civilian Technology', *Research Policy*, March 1974, p.9.
9. *Fortune*, 30 April 1980, pp.12–18.
10. *Frequent European Misrepresentations and/or Misconceptions*, The Boeing Commercial Airplane Company, Oct. 1984, pp.1–2. The Germans do have problems in making large sums from exports of military equipment due to stringent restrictions about to whom they can sell military goods.
11. Rolls Royce, for example, noted that it might be able to sell a 1000 engines of a type to various military projects. GE and P & W could expect to sell between 5 and 15 000.
12. For example, in Britain, the RAF has tended to order small numbers of

transports after they have been built for civil use. An RAF order for the Comet was an important part of the rescue of De Havilland in 1953. See K. Hayward, *Government and British Civil Aerospace*, op.cit., pp.35–7.

13. HoC 389, Q225; HoC 347, 1971–72, Q323.

14. See Roger Chanut, Airbus Industrie Vice President, *European Management Journal*, Vol. 3, No.2, 1985.

15. See the Expenditure Committee's two volume study of public money in the private sector, HoC 347, 1971–72, which examined at some length the reasons why private capital shied away from investing in technologies like civil aerospace. See also, E. Previdi, 'The GATT Agreement on Trade in Civil Aircraft', *Air Transport Economics*, June 1980, p.481. The collaborative dimension, and the legal problems of lending money to collaborative programmes has also been a disadvantage to European industry.

16. *Le Nouvelle Economiste*, 5 March 1984, pp.69–70; *Financial Times*, 31 October 1983.

17. *Guardian*, 10 November 1983; *The Economist*, 27 August 1983, p.22.

18. Boeing, October 1984, pp.3–4.

19. *Financial Times*, 31 October 1983; *AWST*, 27 February 1984, p.35.

20. Hayward, p.183.

21. *Flight*, 5 May 1984, p.1216.

22. *Financial Times*, 31 October 1983.

23. Ibid., 24 April 1985.

24. The French Government, in an attempt to reduce the direct cost of supporting the Airbus proposals are trying to put together a package of Government backed credits. *Financial Times*, 25 February 1986.

25. Airbus Industrie has claimed that the initial sales for the A320, over 100, are a record for an aircraft before first flight, a claim inevitably disputed by Boeing.

26. *Financial Times*, 31 October 1983; *Flight*, 25 October 1980, p.1576; 8 March 1986, p.5; *Interavia*, November 1980, p.1002; *AWST*, 27 February 1984, p.35. One objective calculation suggests that 400 A300s will be delivered before the programme peters out in the early 1990s. See Hugh Cowin, *Flight*, 15 March 1986, p.9.

27. *Le Nouvelle Economiste*, 5 March 1984, p.69.

28. *Financial Times*, 24 March 1985.

29. *Flight*, 11 November 1978.

30. *Flight*, 26 May 1984, p.1380.

31. See Hayward, pp.221–35.

32. The French Comptes de Cours has expressed some concern about Aerospatiale's Airbus account. *Le Monde*, 30 June 1985.

33. J. Newhouse, *The Sporty Game*, op.cit., p.29.

34. See J. Pearce, *Subsidized Export Credit*, London, 1980, ch. 1, and R. E. Feinberg, *Subsidizing Success*, London, 1982, ch. 1.

35. Feinberg, p.131; *The Guardian*, 11 February 1981.

36. See Pearce and Feinberg for a detailed elaboration of the European and American export credit systems.

37. *Guardian*, 11 February 1981.

38. Pearce, op.cit., p.21.

39. J. J. Emery (*et al.*), *The US Exim Bank*, Boulder, 1984, p.10; HoC42, 1963–64, para. 115 & Q678.

40. *Fortune*, 21 April 1980.

41. Newhouse, pp.61–73; *AWST*, 10 April 1978, p.223; *Sunday Times*, 9 April 1978.

42. *AWST*, 13 September 1982, pp.18–19.

43. Newhouse, pp.38–44. For example, the French were prepared to extend credits to the Lebanon based MEA airline in the middle of the civil war. A risk which even the British and Germans were reluctant to accept. *Flight*, 9 July 1983, p.769.

44. In 1984, the Turkish Government decided to buy 7 A310s as a political and economic gesture to the EC states. The deal had seen lobbying of Turkish officials and politicians by Joseph Strauss for Airbus, and ex-Secretary of State Alexander Haig for American industry. Airbus was presented as a German aircraft to circumvent Franco-Turkish trade disputes. Similarly, at a crucial point in the sales campaign, the Turks were offended by a US Congressional attack on Turkish atrocities towards the Armenians in World War I! (*Financial Times*, 23 August 1984). The French also faced similar problems selling aircraft to Australia with disputes over nuclear testing and EC agriculture policy affecting relations between the two countries.

45. *Interavia*, July 1978, p.585; *Flight*, 22 April 1978, p.585.

46. Previdi, op.cit.; J. Rallo, 'The European Communities Industrial Policy Revisited: the Case of Aerospace', *Journal of Common Market Studies*, Mar. 1984, pp.264–5; *Interavia*, July 1979, pp.655–6.

47. Rallo, p.265.

48. *Interavia*, July 1979, p.656.

49. Rallo, p.265.

50. Previdi, p.479.

51. Ibid., p.480.

52. *Interavia*, July 1979, p.656.

53. Ibid.

54. Pearce, op.cit., pp.39, 41–54.

55. In one case, TWA sold tax credits on an L1011 for $13.4 million, or about 30% of the aircraft's value.

56. *AWST*, 12 July 1982, p.30; 26 July 1982, p.30; *Flight*, 10 April 1982,

p.907; 4 September 1982, p.650.

57. *AWST*, 26 June 1981, p.27; 7 June 1982, p.23.
58. *Flight*, 14 November 1981, p.1459.
59. *The Economist*, 8 May 1981, p.90.
60. P. Hartland-Thunberg & M. H. Crawford, *Government Support for Exports*, Lexington 1982, p.67.
61. *The Economist*, 8 August 1982; *Flight*, 26 June 1982, p.1642.
62. *AWST*, 31 May 1982, p.30; 18 October 1982, pp.16–7; 14 February 1983, p.21.
63. *AWST*, 6 September 1982, p.58.
64. *AWST*, 7 June 1982, p.22.
65. *Flight*, 16 October 1982, p.1089.
66. *AWST*, 4 October 1982, p.36; *Flight*, 16 October 1982, p.1089.
67. *AWST*, 16 May 1983, p.21; 8 August 1983, p.23.
68. *AWST*, 28 November 1983, p.22.
69. *Financial Times*, 3 July 1985; *AWST*, 8 July 1985, p.34.
70. *AWST*, 18 March 1985, p.203.
71. See *Interavia*, Jan. 1985, pp.70–1; *Flight*, 1 June 1985, pp.187–9. For details of some of the more extreme examples of selling civil aerospace products, including the aircraft for oil deal with Saudia Arabia, see *Sunday Times*, 5 August 1984; 29 April 1984; *Flight*, 12 May 1984, p.1245; *AWST*, 23 April 1984, pp.31–2, *The Economist*, 5 May 1984, p.68.
72. *Flight*, 9 October 1982, p.1; *Financial Times*, 30 March 1983; *AWST*, 4 October 1982, pp.28–9.
73. *Flight*, 10 March 1984, p.616; *Interavia*, May 1984, pp.447–8.
74. *AWST*, 9 May 1983, p.32; *Flight*, 6 August 1983, p.296.
75. *Interavia*, Jan. 1985, pp.70–1; *Flight*, 1 June 1985, pp.187–9; 1 June 1985, p.188.
76. *AWST*, 30 January 1984, p.45.
77. Panam did order Comet Threes before the fatal crashes in the early 1950s, and placed options on the Concorde.
78. *Flight*, 22 September 1984, p.713; *Guardian*, 14 September 1984.
79. *Financial Times*, 28 July 1984; 17 September 1984; *Sunday Times*, 16 September 1984.
80. *Flight*, 22 September 1984, p.713; *Europe '85*, Jan./Feb., 1985, p.17.
81. *Flight*, 22 September 1984, p.713–14.
82. One of the uncertainties was Panam's need to sort out its labour problems. *Financial Times*, 31 May 1985.
83. *Le Monde*, 15 September 1984.
84. The Air India battle was waged over 18 months, with both sides going for large price discounting. In the event, Airbus Industrie swung the deal on a complex leasing package which effectively reduced the price

of the A320 $25 million, some 25% off its list price. American reaction was particularly incensed; the American Ambassador to India delivered a sharp protest, and Vice President Bush made a personal appeal to Prime Minister Ghandi. Boeing, of course, claimed that Airbus Industrie had made substantial and uncommercial reductions in the price of A320s in order to further encourage a strong current running in favour of the European 150 seat airliner. However, dealing with the Indian Government has always been a delicate and difficult process, and in addition to the terms of Airbus Industrie's offer, wider political factors affected Air India's choice. The final financial package, a mixture of export credits and commercially raised capital, was one of the most complex in civil aviation. *Financial Times*, 2 September 1985, 24 September 1985, 5 March 1986, *Flight*, 15 March 1986, p.4.

85. See note 42, Chapter 4 for details of UDF technology, and *AWST*, 11 March 1985, pp.38–9, 10 June 1985, pp.44, 49; *Interavia*, March 1985, pp.289; June 1985, pp.615–18; *Flight*, 8 June 1985, p.5; 15 June 1985, p.8.
86. *The Economist*, 1 June 1985, p.24.
87. MDD have already signed up Aeritalia, Saab, and the Chinese as partners for its UDF aircraft. Boeing have re-opened negotiations with the Japanese, See Chapter 3.
88. Rolls is, in fact, devoting some 25% of its civil R & D budget, about £3 million to UDF research. Some of this is being provided by the British Government.
89. *Flight*, 13 July 1985, p.5; 1 June 1985, p.80.
90. *Flight*, 15 March 1986, p.4; *Guardian*, 20 January 1986; *AWST*, 10 February 1986, p.38; *Times*, 16 January 1986; *Financial Times*, 27 December 1985.
91. *AWST*, 16 December 1986.
92. Emery, op.cit., p.69; *AWST*, 20 May 1985, p.69; 7 October 1985, p.78; *Financial Times*, 6 December 1985; 10 January 1986.
93. Ibid., 25 January 1986, p.4.
94. *Financial Times*, 16 January 1986.

Concluding Observations

Over the last twenty years, virtually every aerospace manufacturer has adopted collaboration in some form or another as a conventional way of doing business. At the same time, civil aerospace has remained a fiercely competitive industry, where even collaborators on one project can be simultaneously competitors on another. One of the fascinating aspects of this industry has been the way in which companies have managed to live with this apparent contradiction. A key to resolving this inconsistency is to realise that cooperation and competition are not opposite ends of a spectrum, but an interwoven tapestry of relationships. Collaboration is for some a means of reducing the deleterious effects of competition, and for others a way of competing more effectively. Different markets, and differing industrial circumstances can, of course, make it easier for firms to cope with potential conflicts of interest. For the new entrant, cooperation is the only realistic strategy, and for the established firm, it can facilitate a continuing presence in more market sectors than independent production could safely sustain. In a European context, for large civil aircraft at least, it has been a strategic solution to a long-term problem of matching US domination. Even for US producers cooperation, or multinational sourcing, has served to ease the growing difficulties of maintaining a leading position in the industry.

Cooperation is never easy, and on balance, most companies would prefer independent production. For established firms, a proud history of independent enterprise has led to the development of patterns of working which necessarily have to be compromised when working with others. Collaborative decision-making, especially when it involves several governments, can be protracted and often frustrating for industrialists conscious of slipping commercial opportunities. The administration of transnational programmes can also bring extra complications and problems. The memories of past failures and suspicion of long-term motives for cooperation add further difficulties. Certainly, trying to pin down why one collaborative programme has worked and why another has failed is never easy. According to Hochmuth, the basis for a successful transnational venture is a combination of favourable circumstances and an appropriate organisational structure. The former largely depends upon the existence of mutual trust between the partners, which is, in turn founded on a strong sense of common purpose: 'If a joint venture is a "shotgun marriage" between partners, it will not likely succeed even if it appears the right thing to do. No joint venture can for long conceal disparate goals, a set of ulterior motives, or a lack of mutual trust and respect....Once started, self serving actions lead to a vicious circle of mutual distrust that is most difficult to break'. [1] Clearly, the most powerful stimulus for cooperation is a sense of 'live together or die', the perception on the part of potential collaborators that there is little alternative to working together in the face of potentially overwhelming international competition. As Robert Carlson of P & W put it, 'competition is wonderful, but masochistic industrial self-destruction is quite another thing'.

Even this cannot be taken for granted. Hochmuth also considered the 'merry widow' syndrome, where one partner feels that it has achieved enough from cooperation to strike out on its own. [2] This fear is clearly a factor in relations with the Japanese. Both European and American manufacturers are wary of conceding too much to Japanese companies on technology transfer, or in the equally important aspects of sales and marketing. Whether or not Japan has any realistic hope of obtaining an independent capability in large airliners and aero-engines, the belief that cooperation could foster a potential competitor, underlined by a reading of past examples of industrial and technological 'imperialism', has aroused a strong feeling amongst the established aerospace companies which limit the degree to which Japan is wholly welcome as a partner, despite its evident financial attractions. Similarly, the suspicion that collaboration might lead to dependence and a long-term erosion of national technological and commercial skills, can have an equally damaging effect on a prospective partnership. If collaboration is to prosper, it is necessary that the environment for independent production is at the outset, and remains, unpromising. If the temptation to 'break ranks' is low (if only because there are too many uncertainties associated with any alternative course of action) and if the fear of dependence is countered by effective procedures for distributing the technological and commercial rewards of a project, then collaboration has a chance to develop a momentum which may become self-sustaining.

This has happened for the members of Airbus Industrie. If the European airframe industry was to compete effectively with the Americans in the development of large civil aircraft, collaboration was the only realistic route. As one British aerospace executive suggested, Airbus shows that 'where there is the will, a worthwhile product and the right machinery, collaboration succeeds'. [3] But precisely why the Airbus has succeeded remains a mystery; as one European industrialist put it, 'the only guideline is that there are no guidelines'. [4] It may also be that the Airbus does not necessarily stand as an example readily repeatable in other European industrial sectors. [5] As the chequered early history of the programme illustrates, even the survival of the Airbus programme was never certain. Despite the obvious competitive weaknesses of European civil aircraft production, national policies and interests were not always supportive of collaboration.

However, the logic of cooperation continued to be valid long enough decisively to shape the infrastructure of European civil aerospace. From the early 1960s, the Airbus partners recognised that they faced overwhelming international competition, and that they needed to develop an industrial, technical and financial capacity of a scale to match the Americans. In short, European airframe companies had little alternative to cooperation despite its difficulties. The British position is particularly instructive in this respect. HSA and later BAe, despite the withdrawal of governmental assistance, recognised the singular importance of staying in touch with the programme and, later, to resist the temptation of joining American programmes. Similarly, despite evident frustrations with their partners, the Airbus offered the French a more secure future for their industry when faced by similar proposals. The political context has, of course, given the Airbus a particularly powerful source of continuity. Without the support of the German and especially the French Governments, Airbus would not have survived its initial commercial setbacks. In this sense, European industrial collaboration has on balance been well served by its links with wider national interests.

Although state support was, and remains an important element in binding the Airbus partnership together, its success has come to depend upon more positive factors. In Hochmuth's view, the organisation, or structure of cooperation, must continue to satisfy the interests of every member. A successful organisation has to ensure an equity of reward and obligation, and to strike a balance between its own autonomy and the managerial requirements of the participants. This again takes time; and although the choice of a structure at the outset of a collaborative programme can make a considerable difference to the likely success of cooperation, 'achieving a successful trans-national organisation is a matter of years—not months'. [6] Within the Airbus consortium, for example, perceptions of common interest have indeed been reinforced by working practices which have given all concerned equal access to industrial and technological rewards. Airbus Industrie has also had the strength and flexibility to overcome a number of difficulties— differences of opinion over management style and corporate objectives—the allocation of subcontracts—and the need to adjust the national balance of top management in line with changing national contributions. The Airbus partners have also developed that all-important sense of trust. They have found that cooperation works, and that they can build on earlier experience to produce an adaptable and evolving collaborative structure. The continuity of development may well have produced a qualitatively different form of European collaboration where many of the past criticisms of joint ventures may be much less valid. Airbus Industrie may face problems in the future, especially in launching its latest projects without substantially disturbing the existing patterns of work sharing and decision making; but given the determination of the French and the German, and perhaps the British, Governments, to maintain Airbus as a strategic asset, these problems will assuredly be overcome.

Airbus was also well served by a number of individuals, most notably its first two senior managers, Bernard Lathière and Roger Béteille, whose unremitting advocacy of the aircraft was crucial both in establishing commercial credibility and mobilising political support on behalf of the programme. Equally, the interlocking connections of Franz Josef Strauss cannot be underestimated in confirming German commitment to the project. Similarly, the interchange of French personnel between the consortium, Aerospatiale and the French civil service, has been a consistent source of support. Even in the United Kingdom, the belief of individuals in the importance of the project to the British aircraft industry served to maintain British interest even when government support had waned and the pressure to seek alternative solutions was substantial.

Cooperation between the American and the more important European airframe companies, on the other hand, never really overcame major differences in outlook and industrial interests. Although the pressure to share costs and to expand market access was felt by American firms, the main stimuli for cooperation were largely defensive. Their actions may not have been a deliberate attempt to undermine the Airbus consortium, but clearly to have detached a major European aircraft industry would have had a radical effect on the structure of international competition and cooperation. In practice, neither Boeing nor MDD was prepared to concede enough on important managerial and technical issues to overcome deep-rooted European suspicions. There was a feeling that the Americans could not overcome the perhaps unfounded fear in Europe that they wanted to turn European companies into mere 'metal bashers'. Collaboration was never sufficiently vital to force significant concessions from the Americans. These elements, however, were paramount to the major European firms, and were regarded as being integral features of the reward structure of collaboration. If the Americans seemed to be making progress it was again largely for negative reasons. Dissatisfaction with the progress of European cooperation, or, as in Fokker's case, fear of overdependence on Airbus, were more important motives than the intrinsic value of collaboration with US companies. The negotiations between American firms and the Europeans never generated a satisfactory degree of empathy. This was undoubtedly due in part to the different philosophies of industrial development held by Europeans and Americans. In the American view, their hard-nosed attitude to commercial decision-making and central control was ill-suited to the existence of wider social and political interests underlying European industrial activity.

In recent years, even though Boeing has shown some willingness to make more concessions in order to reach collaborative agreements, notably with the Japanese, there are still limits to how far the American companies are prepared to compromise their central, decisive role in project management. In the final analysis, the Americans do not need a 'politically' inspired form of collaboration. Unlike the European and the Japanese position, cooperation for Boeing, although perhaps increasingly less so for MDD, can be viewed as one element in a corporate strategy firmly based on an independent capability. If they feel the commercial potential justifies the risk, American companies can and will launch programmes independently, and only then seek to attract foreign sub-contractors or risk-sharing partners. Collaboration has been incorporated into Boeing's strategy of developing a 'family' of civil aircraft, and the company has also extended its range by acquiring De Havilland of Canada. This will enable Boeing to offer a range of commuter aircraft to the smaller airlines which, in the US at least, are establishing links with Boeing's more traditional customers. It has also reached a preliminary agreement with MBB and the Indonesian aircraft industry to develop an 80 seat prop-fan airliner. These developments, combined with a further extension of the 747 range, mean that Boeing will be unique in the world's aircraft industry in its ability to offer a family of products from 30 seats to 500. [7]

Airbus Industrie's threat to Boeing lies precisely in its goal of matching Boeing's products with a 'family' of its own. This has been a consistent objective from the 1970s, and underlies the pressure to launch the A330/340. Pierson has described these two projects as 'the last large investment' that it will request from its member companies (and by implication from their sponsoring governments), to achieve 'a strategy of offering a complete family of aircraft'. [8] The existence of several European projects, some developed by Airbus Industrie's members, in the smaller aircraft category, will obviate downward expansion of the Airbus family. However, taken as a whole, Airbus and these projects will ensure that Europe does have a complete range of civil aircraft. MDD, although it will offer competitive products in a number of markets, is unlikely to present such a comprehensive challenge. Airbus' presence, so firmly backed by European governments, obstructs Boeing's ability to control the market and to derive the commercial advantages of monopoly.

Although the competition between Airbus and the Americans does tend to dominate views of the civil aircraft industry, the pattern of competition and cooperation in the engine sector has taken a rather different form. The political history of European civil engine development was a decisive factor in stimulating a trans-Atlantic orientation to cooperation. There was little empathy between Rolls Royce and Snecma; indeed, the reverse would probably be a more accurate description of their relationship. The two companies had different points of departure; for Rolls it was the protection of an existing capability and technological status; for Snecma it was a question of making an industrial and commercial breakthrough. For the French, collaboration was an essential feature of their strategy; for the British it was an element, albeit a vital one, in completing a range of competitive products. In this sense, Rolls' position more closely resembled that of GE and P & W, although GE, with less experience in civil engines, has tended to embrace collaboration more positively than either P & W or Rolls.

In general, cooperation in the engine sector has been more like an *à la carte* formula than the comprehensive format of Airbus Industrie. Although Snecma and GE are moving towards collaboration on the next generation of civil engines, and incremental developments have been accommodated within CFM and IAE, these partnerships are limited to specific projects aimed at particular market categories. The GE–Rolls link is somewhat more flexible, and involves two different projects, but it has yet to lead to a joint programme. Cooperation is still, and is likely to remain, based on discrete projects designed for specific market categories, and the industry as a whole is quite firmly based on national companies. The organisation of cooperation in civil engines has helped to maintain the separation of interests, involving the exchange of components rather than technology. The 'natural' divisions of engine construction has allowed each partner to concentrate on an aspect of engine design and development without requiring comprehensive agreement on technology transfer. This may also have helped the engine sector more readily to sustain simultaneous competition and cooperation between the major companies. The three major companies are able to protect areas of expertise, yet contribute to the success of a joint programme. The establishment of separate organisations to coordinate development, production and, especially, the commercial exploitation of a joint product has made an important contribution to the success of IAE and CFM International, but they have not implied a permanent set of industrial linkages. This fact has also limited the danger for the established companies of encouraging new sources of competition from the dependent companies. While the smaller firms, and technically this includes Snecma, have been able to share in the commercial benefits of collaboration, they have not obtained the degree of technical reward which they might have expected from cooperation, and which would be necessary to ease the undoubted cost of establishing an independent capability.

Interestingly enough, the engine companies have been relatively quiet about the broader politics of the international civil aerospace industry. Undoubtedly, the fact that their relations do not conform to a simple pattern of transatlantic competition has had an influence on their position. Whatever the outcome of battles in the airframe sector, airliners will still need engines, and to a degree, the more aircraft on offer, the more opportunities the engine companies have to sell their products. They might be affected by the onset of protectionism and any other measures designed to emphasise national sourcing of equipment, but the complexity of their mutual links would moderate the effects of any deterioration in relations between the United States and Europe. The Europeans, and especially Rolls Royce might be more vulnerable to protectionist measures, being far more dependent on the American market than P & W or GE are on the European. But clearly, neither the American or the European engine manufacturers can be sanguine about the possible consequences of a more serious conflict between the United States and Europe over high technology trade. They must certainly be concerned about the strident tone of the US–European debate about civil aerospace and other high technologies.

In general, civil aerospace is an area of advanced technology where, on technical merit alone, there has never been a 'gap' between Europe and the United States. Even the much criticised Concorde project is an operational tribute to the excellence of European design and technological expertise. Similarly, in technical terms, Airbus is outstandingly successful, and in many respects, Airbus Industrie believes that it has (at least for the present) wrested the technological initiative from the Americans. As Adam Brown of Airbus Industrie put it, 'the tables have turned when it comes to new technology'. [9] But Concorde is a double-edged symbol. The one vote defeat of the American SST in the

Senate was to many in the United States a victory for common sense and commercial logic over politically inspired technology. In the same way, despite its evident quality as a product, Airbus' commercial standing is still debatable. If effectiveness is measured by profitability, the project has still some way to go before it can be regarded as a viable enterprise. The bottom line of success in civil aerospace must be commercial return. As Sir Arnold Hall of HSA put it, 'there's only one eventual way of judging the viability of an aeroplane, just as on any other commercial product, and that is, "Does the profit and loss account look right?"' [10] Airbus has undoubtedly made a market penetration that hitherto Europeans had only dreamed of, but it has been an expensive achievement. Although a return is being made on public money invested in the programme, clearly the first two projects must be seen as an investment to establish Airbus as a credible force in the market. The A320 and subsequent projects will have to be judged on stricter financial criteria if Airbus is ultimately to be counted as a commercial success.

However, Airbus has to be seen as being more than a simple commercial exercise. It is a strategic challenge to the United States, and even if this philosophy has not been uniformly accepted by all its industrial and political actors, this remains a fundamental rationale of the Airbus programme. That being said, the pressure to cut costs and to maximise industrial efficiency within the constraints of a multinational enterprise in order to improve the project's commercial standing, is evident throughout the programme. But from a political standpoint, the continuing presence of European industry in the market for large airliners, and civil engines for that matter, may well be the most important consequence of the past twenty years of effort by Airbus Industrie, as well as by Rolls and Snecma. In the final analysis, that has been a consequence of governmental commitment and national will.

The French have been the most consistent advocates of building a European civil industry capable of challenging the Americans. They have been prepared to pay the price of beating Boeing, or at least to help the European civil aircraft industry survive in the 'shark's pool'. In this respect, the development of civil aircraft and engines has been part of a wider strategy designed to achieve a degree of independence in key technologies. They are also confident that the wider implications of Europe's efforts for better relations with the United States can be contained. Some French officials and politicians tend to dismiss Boeing's complaints and the increasingly hostile noises from Congress as the natural reaction of a special interest lobby to unaccustomed competition. If the Americans have benefited in the past from market rules which have been biased in their favour, they will simply have to get used to a new set of conditions in which European governments actively intervene to redress this situation. Nevertheless, even the French recognise that overt protectionism could damage their interests. In any event, civil aerospace has been targeted as any area where, under French leadership, the United States will not be allowed a controlling influence over European industry and technology. There may be a degree of inconsistency where aero-engines are concerned, but as in satellite launchers, the French will not allow the technological initiative to remain on the other side of the Atlantic. There is, as Regis Debray put it, a determination to protect Europe from 'une progressive satellisation sur l'orbite américaine'. [11]

The German view has been to stress the more positive side of Europeanism, especially the concrete realisation, in projects like Airbus, of the Franco-German relationship and the political rehabilitation of West Germany. Traditionally, their links with the United States since the war have made Germans unhappy with actions that tend to set Europe against its American ally. Nevertheless, the determination to achieve national industrial and technological goals through collaboration has given the Germans a vested interest in the success of European programmes. Increasingly, the Germans have begun to assert themselves in European programmes like Airbus. Equally, in the case of the EFA project and in their reluctance to support the Hermes reusable space vehicle, German policy may not necessarily mean an unquestioning acceptance of French leadership. Yet, and here the British must continually recognise a basic reality, the Franco-German axis will continue to play a decisive role in shaping Europe's technology policies.

British reluctance to embrace European 'grand designs' has been in evidence throughout the Airbus programme, often reflecting the divergent interests and attitudes of its aircraft and engine industries. The British have been less inclined to see civil aircraft development in terms of a crusade against the Americans. Indeed, in other areas of aerospace, there has been a strong interest in collaborating with the Americans. This has often led the French in particular to question where British loyalties really lie. Yet as the furore generated by the Westland affair and by the various American bids for the remaining elements of the motor industry in British ownership, have shown, there is growing apprehension in the UK about the danger of further American control over vital technological and industrial assets. This, as one commentator put it, may reflect more a nascent anti-Americanism than a full-blooded commitment to European solutions, 'a mean-minded sentiment, based on the resentment of the weak for the strong, the poor for the rich, the ossified for the dynamic'. [12] Nevertheless, the pressure of events, and the contradictory strands of British policy towards both the United States and Europe have again conspired to expose the lack of clear direction in industrial and technological issues.

The Westland debate highlighted a growing scepticism amongst some British politicians about an American definition of partnership, while at the same time underlining the lukewarm Europeanism of the Thatcher Government. Paradoxically, the Westland company defended a clear commitment to an extension of a long established partnership with Sikorsky, and an equally clear suspicion of the motives of some of its erstwhile European rescuers. But, as Michael Heseltine put it, the issue was the relationship with the United States in key technologies. If there was any strong sense of a growing imbalance in the partnership, the 'political process would be uncontrollable'. The US had the potential to 'buy its way through sector after sector of the world's advanced technologies', a power born of its huge domestic market and implicit government protection. According to Heseltine, it was in the interest of both the US and Europe that the latter should 'strengthen and coordinate' its industrial base. [13] The House of Commons was by no means united in its support for Heseltine's stark view of American motives, but he did receive the backing of James Callaghan, Prime Minister at the time of Boeing's negotiations with BAe in 1978. Showing a good politician's selective memory, he noted how BAe and his Government had resisted a 'prowling' Boeing, set on sucking BAe dry of technology. The result, he said, was further British participation in the Airbus, and a 'moral for helicopter production—that Europe has an independent aircraft industry'. [14]

The British Government did not come out of the Westland affair with much credit for forward thinking or for having any clear concept of industrial policy. In the short term, the British probably got little credit from either the Europeans or the Americans. According to one French industrialist, 'if ever Britain shows an upsurge of dignity—a refusal to become the vassal of the Americans—then we will accept them.' From an American perspective, the 'exclusionist' sentiments heard from some British politicians were the worst since de Gaulle and Giscard. [15] The British do genuinely find themselves in a cleft stick. Already US investment in British industry is greater than in any other foreign country after Canada. Equally, American multinationals have been 'model corporate citizens', with a solid record on R & D, export earning and local employment of senior personnel. [16] Moreover, British officials and politicians are less sanguine than the French about American reaction to the European technological challenge. The British are worried that the pressures for an active American response to lost markets and a diminishing technological leadership may become a major threat to their, and Europe's wider interests.

The general climate in Western trade, industrial and technological relations does remain tense, and subject to numerous sources of irritation, suspicion and political concern. Although the return of more buoyant market conditions has taken some of the edge off more aggressive American attitudes, pressure for an active riposte to European 'subsidised competition' has not disappeared, particularly as Europe has a general trading surplus with the United States. The possibility of a more dynamic United States policy to maintain and to promote its commercial aircraft industry certainly remains on the agenda of American public debate. Indeed, Leon Brittan, one of the political casualties of the Westland affair, warned that outbursts of anti-Americanism would not help to defend Airbus against its growing band of American critics. [17] The lines of debate in the United States are now distinct. The economic 'right', President Reagan's Council of Economic Advisors, for example, has advanced a hard line, anti-interventionist philosophy. Export and other subsidies should be left entirely to the Europeans and the Japanese; if they want to subsidise civil aeronautical research and sales, that is their choice. If it led to cheaper aircraft for American airlines, and lower fares for

American passengers, this would be to the advantage of the American economy. [18] On the other side, the Keyworth Committee of the Office of Science and Technology came out strongly in favour of a more overtly interventionist strategy to promote all aspects of United States aerospace. The Keyworth Report echoed long established European beliefs about the wider economic, technological and strategic importance of aerospace. In particular, it stressed the vital importance of a competitive US civil aircraft industry. Through its Aerospace Policy Review Committee, the OST has continued to press for a more active approach to high technology trade; 'there are nations in the world catching up to us. To keep up, we need to maintain an environment where the development of civil aviation can flourish and to ensure the United States aerospace industry a technical base. It is imperative for the US to maintain pre-eminence'. [19]

The flow of advice from the Keyworth Committee could be seen as a reflection of its membership, many of whom are ex-employees of the US aircraft industry. Nevertheless, its views are consistent with other statements from business groups, *ad hoc* committees and from individual members of the US Congress. The President's Commission on Industrial Competitiveness, another 'blue riband' group, put forward a general case for improving US industrial performance. It too advocated a more active role for government. It certainly wanted to see relaxation of internal restrictions on US industry, to 'reflect the reality of international competition'—to allow, for example, anti-trust exemptions 'for mergers that promote national objectives'. [20] Even moderate economic liberals, who see 'counter subsidies' as a 'second best alternative' to a free market in high technology trade, have argued that they are justified as a 'safety net to counter low-level foreign subsidies to competing industries abroad'. According to this view, the wider implications of a healthy R & D base are obvious. It implies a 'sustained world demand for the exports of the country, an internationally strong currency, and enhanced power and prestige in the international arena'. Further erosion of American high technological capabilities could reach the stage where the economy 'harbored as many weaknesses as strengths. A trend in that direction, if not reversed, would invite decay and, with it, a suggestion of second class'. [21] Some American industrialists feel that the commercial spin-off from large scale defence programmes like the Strategic Defence Initiative, will not be sufficient without active government direction to stimulate commercial innovation. [22] A hint, perhaps, that defence spending alone cannot generate the technological innovations needed to sustain American industry at large into the next century.

Clearly, Washington politics follow their own rules, and an Administration speaks with many bureaucratic voices. For every piece of advice there is a counter in the classic pluralist tradition. There is still a deep commitment in American business, government and in the body politic generally to free market policies, 'the instinctive style of the US Government is to leave well alone'. [23] However, in the past, crises in individual industries have led to massive Federal intervention, and as American structural advantages are more deeply undermined by active European and Japanese policies, the pressure on the US Government to do more will surely grow. Even where private companies themselves may wish to keep government out of their affairs, 'as their sense of vulnerability to influences beyond their control...increases, so attitudes towards government protection have mellowed'. [24] For the moment, the American civil aircraft industry for one does not want direct subsidisation. But what it does want is a return to the conditions which favoured the natural advantages of the American producer. Given the commitment of Europe to civil aerospace, this is unlikely to be so easily managed. There are hints of a more active US response to the European challenge. Reagan's free market, free trade instincts have been qualified by a modest, Exim administered 'war chest' to support American exports. There are proposals wending their way through Congress to reform US anti-trust law to make internal cooperation easier in the face of international competition. Equally, Treasury Secretary Baker's plan for concerted intervention and coordinated bailing out is an admission that free markets have failed. Taking its text from the American *Business Week International*'s plea for a reconsideration of government attitudes towards interventionism, one British newspaper suggested that, 'in a way the conceptual leap towards an industrial policy has already been made'. The themes of European pundits such as Michael Heseltine are also the themes of American observers and 'increasingly, American politicians'. [25]

One powerful voice in favour of a more active US industrial and technology strategy is that of Lee Iacocca, the hero of the Government led Chrysler rescue. Surveying the threat to American industry, he warned, 'I don't know when we're going to wake up, but I hope it's soon. Otherwise, within a few years our economic arsenal is going to consist of little more than drive-in banks, hamburger joints, and video game arcades'. Although he is a fervent advocate of the American capitalist system, his experience at Chrysler led him to realise a need to qualify ideology with good sense. The ideologues of the free market, he suggested, want America to be great again, 'but by accident'. The plain truth was that 'the market place isn't always efficient. We live in a complex world. Every now and then the pump has to be primed'. Planning was not un-American, it was an essential feature of corporate life and of American football. It was only un-American when government attempted to intervene in the mysteries of the market place, and 'because of this...we're the only advanced industrial country in the world without an industrial policy'. Iacocca opposed government intervention for its own sake, and too much involvement by the state in corporate life, but as the US continued to fall behind its main competitors, it was because 'it's become increasingly clear that there's also such a thing as too little.' [26] The future may see a resolution of present American policy contradictions, if not by the current government, perhaps on the part of a successor. We may, then, have reached, or be about to reach a threshold in US industrial and technological policy which will have massive ramifications for the western industrial system generally, and for civil aerospace in particular. Even a partial move towards European style interventionism, a TVA for American industry as Iacocca might put it, could focus all the inherent power and energy of the United States on key technologies to such an extent that the European 'challenge' would be swamped, and the success of projects like Airbus, seem small beer and all too fragile by comparison.

In Europe the perspective is naturally rather different. The statistics of US market dominance, even in commercial aerospace where the Airbus has made substantial progress, or where Europeans and others share in US success, do little to still fears of a continued '*Défi Américain*'. On a wider technological front, in spite of recent sales to the United States, the two-way street in defence equipment seems uncommonly free of east to west traffic; restrictions on technology trade to the Communist world seem to constrain non-American firms rather more than those in the United States; and US guarantees about technology transfer if Europeans collaborate in

SDI projects and other areas of common concern, have been greeted with considerable caution. The 'Eureka' initiative, even though at the moment it may ask more questions of European technology than it provides answers, is based on the assumption that a collective, coordinated response is needed in the face of a threat that a huge, state-sponsored technological surge under the SDI umbrella could overwhelm European industry across a wide spectrum of technologies. The overall perception is still of American superiority in advanced technology industries stemming from fundamental structural advantages. In civil aerospace and similar sectors the United States still commands the industrial high ground. As Arthur Reed put it, 'even if some of its traditional foreign markets are eroded, there is no question that the US industry, with its immense capability for both military and civil products, will remain as far as one can see ahead the dominant force in world aerospace'. [27] Under these circumstances, European states feel that they must seek compensation through directed industrial and technological policies, and do not accept that Americans should feel threatened by their achievement of a substantial, but not overwhelming share of the market for civil aerospace and other high technology products.

The danger is that the international complexity of the civil aerospace industry, as well as relationships in other industries and technologies, might be forgotten in the clamour to defend national or regional interests. The competition between Airbus and the Americans sometimes obscures the symbiotic relationship between national aircraft industries and individual companies which transcends a simple bipolar perspective. This book has largely concentrated on the 'big' programmes, but even so, the depth of mutual participation in aircraft and engine projects has been continually noted. As such, the international basis of the industry is vulnerable to ill-considered actions taken to defend one element, or even a single company, by imposing restrictive measures on the operation of the industry. The United States may well have the least to lose by either unleashing protectionist sentiment, or by taking active steps to link its vast reserves of financial and technological strength to out-subsidise the opposition. Its industry would survive the ordeal, even if not perhaps as happily as some might contend. [28] Europe, on the other hand, would be the obvious loser. Its best claim is even now to a minority share of the market for airliners and engines. If the United States was to take drastic action to curtail the 'European challenge', the costs of maintaining civil aerospace and other similarly supported technologies would increase dramatically, and the pressure to cut increasingly expensive programmes would rise accordingly.

Looking to the future, the options facing both the United States and Europe are increasingly limited if a downward spiral of conflict over trade in civil aerospace and other high technology is to be avoided. Any ill-judged intervention to redress perceived, but fundamentally erroneous threats to a 'free market' carries a clear danger of undermining the stability of an industrial structure of great value to many members of the Western industrial system. There has been no natural equilibrium in a market place where the state has played such an active role either directly or indirectly in determining its structure and operation. Simplistic demands for a return to a free market in civil aerospace have all the hallmarks of a yearning for a mythical 'golden age'. The Americans might have to bite on the bullet of a competition which could be described as 'unfair', but which does contain a legitimate element of compensation for their implicit advantages in the market. By the same token, a less overtly aggressive posture from some Europeans would be less provocative and more conducive to producing agreement, if only to differ, on the role of state support in high technology. It is, perhaps, too easy for a European to call on the United States to accept the responsibility for maintaining a new status quo, especially when it might imply some costs to American industry. However, the United States has derived considerable advantage from the past balance of industrial power. If it still values the wider benefits of the Western industrial system, this might be an area where concessions could generate a more positive response from Europe on a wide range of other issues.

Notes

1. M. S. Hochmuth, *Organising the Transnational*, Leiden, 1974, p.15.
2. Ibid., p.17.
3. Sir Peter Fletcher, *Sunday Times*, 12 January 1986.
4. Ibid.
5. There are a number of other collaborative projects based on the GIE system, including Euromissile. In June 1985, the Council of Ministers

of the EC agreed to a statute establishing the European Economic Interest Grouping (EEIG) to provide Community status for companies from more than one EC country. This will require member states to adapt national legislation allowing EEIG ventures to be registered. This will take effect from July 1st 1989. Like the GIE, the EEIG is based on the concept of an unlimited liability partnership. An EEIG would not be able to seek investment from the public, but would be a contract between parent companies, registered in any EC state. The EEIG is viewed as a partial solution to the problem of creating a European regime for transnational operation, company law and mergers. The idea had, in fact, been on the EC agenda since 1973. Paradoxically, projects like the Airbus would not come under the provisions of the new EEIG, as it is limited to ventures employing less than 500 people. Its main aim is not manufacturing, but to provide a legal framework for, inter alia, joint research or combined accounting facilities. *Financial Times*, 28 June 1985. Joan Pearce and John Sutton have also argued that the Airbus may not be a model for other European high-technology ventures. See J. Pearce and J. Sutton, *Protection and Industrial Policy in Europe*, London, 1986.

6. Hochmuth, p.18.
7. *AWST*, 9 December 1985, p.28; *Financial Times*, 14 November 1986.
8. Ibid., 24 February 1986, p.45.
9. Ibid., 10 March 1986, p.229.
10. P. Pargamenta, *All Our Working Lives*, London, 1984, p.98.
11. R. Debray, *Les Empires contre l'Europe*, Paris, 1985, p.22.
12. Ian Davidson, *Financial Times*, 20 January 1986.
13. *Hansard*, Vol.89, 15 January 1986, cols 1100–1.
14. Ibid., cols 1112–13.
15. *Financial Times*, 27 February 1986; *Guardian*, 27 January 1986.
16. See for example, a number of articles in the *Financial Times*, 8 February 1986, 20 February 1986. For a more extensive and balanced analysis of US multinational activity in the UK, see J. M. Stopford and L. Turner, *Britain and the Multinationals*, London, 1985, chs. 6 and 7. Shorts' decision to become a risk sharing partner in the Boeing 7J7, with an investment of $100 million, will further underline the ambiguities of Britain's position vis a vis the United States and Europe.
17. *Hansard*, op.cit., col.1165.
18. *AWST*, 7 February 1983, p.34.
19. *AWST*, 3 September 1984.
20. *Financial Times*, 14 February 1985.
21. Hartland-Thunberg, op.cit., pp.85–9.
22. A major barrier to the effective use of defence derived innovation was

the growing length of the US procurement cycle. In the late 1940s and 1950s, when the US economy did derive considerable benefit from the surge in military R & D, the cycle was between four and five years. Now it was 11 to 12 years, and as a result, US industry tended to lose much of the commercial advantage that it once gained from Pentagon inspired activity. The system of effecting technology transfer also lacked efficiency, and anti-trust laws were again cited as barriers against effective exploitation of complex and expensive technology. *AWST*, 25 November 1985, pp.75–9.

23. Edmonds, 'Market Ideology and Corporate Power', op.cit., p.75.
24. Ibid., p.100.
25. *Guardian*, 4 March 1986.
26. L. Iacocca, *Iacocca*, New York, 1984, pp.336–46.
27. *The Times*, 6 November 1985.
28. The Aerospace Industries Association of America has adopted a firmly anti-protectionist stance. But there is still considerable pressure for the US Government to take some active measures designed either to curb European malpractice, and/or to support US civil R & D more strongly. See also, *Flight*, 29 March 1986, p.6.

Index

A300 41, 52, 56, 57, 73, 74, 94,
        110, 171
    A310 and 55, 76, 96
    costs of 58, 163, 165–4
    development of 53–4
    engines for 38, 53–4, 148
    equipment for 77
    significance of 63
A310 41, 52, 74, 97, 110, 148, 183
    A300 and 55, 73, 96
    BAe support for 56–7, 103–6
    costs of 163, 165
    development of 55–8
    equipment for 76–7
    significance of 57, 58
A320 73, 82, 85, 115, 116, 185
    costs of 163–5
    development of 58–63, 69
    engines for 146–7
    equipment and 76, 77
    Italy and 96
    Japan and 114
    Panam order 72, 183–4
    significance of 52, 63, 286
A330 85, 86, 117, 166, 186, 201
A340 85, 86, 117, 166, 186, 201
Action Programme for European
        Civil Aerospace *see* Spinelli
        Report
advancés rèimbursable 38
Aeritalia 40, 95–6, 97
Aerospatiale 13, 118, 160, 197
    A310 and 97–9
    A320 and 58–63
    Airbus R&D 74–6
    Airbus Industrie and 52, 66–7,
        73, 101, 164–6

Boeing and 97, 98–100
Dassault and 97, 98–100, 101
European collaboration and 96
French aircraft industry and
        52–3, 66–7, 73, 85, 101,
        164–5, 166
Airbus *see* Airbus Industrie
Airbus Industrie 38, 40, 43, 98,
        101–2, 104, 161, 202, 203,
        207
    Boeing and 30, 95, 97, 99–100,
        163, 165, 176, 183, 185, 187
    British suspicion of 102–3
    composition of 13
    decision-making, industrial
        66–8
    decision-making, official 69–70
    efficiency of 80–1, 83–4
    equipment contracts and 75–8
    Executive Agency 70
    Executive Committee 69
    export credits and 171–3, 175,
        177
    Financial Committee 60
    financing of 78–81
    Fokker and 110–11, 112
    formation of 54, 65
    framework agreements and
        69–70
    General Assembly 66
    GIE and 54, 65, 78–9, 81, 84
    inter-government agreements
        and 70
    Inter-governmental Committee
        69–70
    Japan and 114
    leasing and 182–3

Airbus Industrie (cont.)
　Ministerial Meeting 69
　nationalities and 81–3
　Panam sale to 183–4
　production system 71–2
　production and sales matching
　　and 72–3
　R&D and 74–5
　reorganisation of 81–5
　sales penetration of 22, 25, 52,
　　171–3, 181, 183–5
　significance of 13, 52–3, 85–6,
　　119–20, 166–7, 201
　sourcing policy 75–8
　strategy 55, 198
　sub-contracting and 75–8
　subsidies and 30, 162–8
　success, reasons for 195
　Supervisory Board 67
　US input to 14, 77
　work sharing 73–5
　see also A300, A310, A320, A330,
　　A340
Aircraft Agreement see GATT
Air France 125–6
Air India 55, 185
airlines 19–22, 25, 54–5, 159,
　　172–3, 174
American Airlines 167, 182
Ansett 183
anti-trust 31, 138, 146, 149, 151,
　　296
ASTR 98
ATR42 84
Australia 44
Aviation Week and Space Technology
　　180

B10 see A310
BAC 34, 54, 97, 104
Bacher, Thomas 1, 2, 47, 187
BAe 146　34, 84
Baker, James 206
Barbré, Yvres 100
Barre, Raymond 101
Baudis, Dominique 184
BEA see British Airways
Belairbus 65

Belgium 40
Berrill, Sir Kenneth 106
Beswick, Lord 97, 106
Betéille, Roger 60, 74, 82, 171, 196
Boeing 13, 22, 30, 34, 55, 80, 94,
　　140–1, 176
　Airbus and 163–6, 172–3, 183–4
　Aeritalia and 95–6
　Aerospatiale and 97–9, 101–2
　BAe and 56–7, 102–6, 203
　collaboration and 94, 112–3,
　　118–9, 120, 197–8
　derivative aircraft and 58, 112,
　　115
　Fokker and 111
　government support and 14,
　　160, 161, 184
　Japan and 47, 113–8, 119
　UDF and 185–7
　US aircraft industry and 27–31
Boeing 367–80 19, 157
Boeing 7–7 115–7
Boeing 707 19, 27, 29, 94, 157, 160
Boeing 727 27, 94, 157
Boeing 737 27, 60, 112, 157
Boeing 737–300 28, 58, 115, 116,
　　135, 182
Boeing 747 27–8, 29, 95, 138, 148,
　　153, 159, 167, 198
Boeing 757 28, 56, 94, 103–4, 111,
　　125, 198
Boeing 767 28, 94, 111, 113–14
Boeing 7J7 117–8, 185, 186
Boeing 7N7 94, 97–8
Boeing 7X7 94, 97–8
Boeing KC135 19, 135, 158, 160
Bouillon, 'Tex' 105
Bradley, Frank 25
Breguet 37
Bristol Siddeley Engines (BSE) 37,
　　128–30
Britain see United Kingdom
British Aerospace (BAe) 97, 10,
　　118, 160, 169, 195
　A310 and 55–7, 103–6
　A320 and 58–63, 73–4
　Airbus Industrie and 13, 40, 52,
　　66, 67, 81, 85–6, 164–5

Boeing and 103–6
British aircraft industry and 34
privatisation and 36
British Airways (BA) 56–8, 60,
102, 103, 182
British Caledonian (B-Cal) 60
Brittan, Leon 204
Brizendine, John 31
Brown, Adam 200
Business Week International 206

C5A 29, 158–9
Callaghan, James 105, 171–2, 203
Carlson, Robert 144
Carter, Jimmy 175, 177
CASA 13, 66, 68, 164
Cavaille, Marcel 99
Central Policy Review Staff
(CPRS) 106
CF6 29, 38, 54, 128, 130, 136, 137,
148, 149, 153, 159
CFM56 29, 38, 127, 139, 144, 152,
186
A320 and 58, 114
French Plan and 38
Mercure and 98, 99
significance of 135–6
V2500 and 146–7
see also CFM International
CFM International 11, 138, 140,
145, 152
formation of 130–5
organisation of 133–5
significance of 135–6, 199
success, reasons for 136–7
Chanut, Roger 71
China 27, 44
Chirac, Jacques 97
Chrysler 207
COFACE 169
collaboration 13–14, 118–20
Anglo-French 39–40
EC and 31–3
engines and 125–7, 150–3, 199–
200
France and 37–40, 62, 128–31,
202–3
Germany and 42–3, 62, 202–3

Holland and 43–4
Japan and 46–7
success of 193–7
United Kingdom and 35, 39–40,
128–31
United States and 31, 93–5, 197
see also Airbus Industrie; CFM
International; International
Aero-engines
Comet 19
Commerce, Department of 77, 132,
179, 181, 183
Commission on Industrial
Competitiveness 205
Commonline 176–81, 183
competition 13–15, 199
Boeing and Airbus and 187–8
engines and 125–7, 134, 141,
150–3
Panam, Airbus sale and 183–5
V2500 and CFM 56 and 146–7
Concorde 20, 37, 38, 61, 64, 76,
128–9, 162, 200
Congress, United States 10, 173,
178, 179–80, 187, 201, 206
Convair 880/890 29
costs, civil aerospace and 1, 18,
27, 32, 41, 58, 112, 126,
162–4, 166
Council of Economic Advisors 204
currency fluctuation 81, 134, 165

Dassault 36–7
Aerospatiale and 97–103
MDD and 97–103
DC-8 19, 29, 158
DC-9 99, 158
DC-10 20, 28, 29, 55, 159
Debray, Regis 202
defence, civil aerospace and 1, 24,
157–60
Defense Production Act *see* 'V'
Loan
De Havilland Canada 198
Delta 59, 182
Department of Trade and Industry
(DTI) 60
de-regulation, US airlines of 25

Deutsche Airbus 53, 62, 64–5, 66, 67, 164
Douglas 25, 158
Draper, William 178, 179
Duncan, Sir William 148, 149

Eastern Airlines 171, 172
ECGD 169, 171
Economist, The 47, 148
EFA 202
Electra 27
equipment industry, European 74–8
Eureka 12, 208
Euromissile 43
Europe
    Airbus and 52–3, 63, 83, 86
    civil aerospace cooperation and 31–2, 95, 101, 106–7, 118–20, 193
    civil aerospace problems of 22–5
    engine industry and 128–9, 153
    export credits and 170–81
    high technology and 8–11
European Community (EC) 31–3, 96, 174, 186
Exim Bank (Exim) 169, 170, 175, 176, 178, 179–80, 181, 187
export control legislation 77, 146, 187
export credits 70, 168–81

F16 109
F27 107, 110
F28 107, 108, 109, 110, 112
F50 44
F100 44, 112
F101 'hot core' 132
FAA 180
Fagard, Oliver 134, 135
family concept 18, 55, 63, 126, 198
Federal Express 159
Fiat 35, 127, 138, 139, 144, 149, 150
Fitterman, Charles 59, 63
Fitzgerald, Ernest 158

FJR 710/60 engine 142
Flosdorf, Heribert 83
Flight International 149, 164, 186
Foche, René 64
Fokker 34, 37, 40, 118, 197
    Airbus and 65, 110, 111–15
    Dutch aircraft industry and 43–4
    Japan and 114–15
    MDD and, 111–12, 114–15
    VFW-Fokker and 107–11
Fortune 160
France 1, 104, 106
    A300 and 52–5
    A310 and 55–7
    A320 and 58–63
    Airbus Industrie and 64–5, 70, 73–4, 76, 81–2, 85
    aircraft industry and 36–9
    civil aerospace, government support for 38–9, 162
    engine industry and 128–37, 147, 150–1
    European challenge and 201–2, 204
    export credits and 169–71, 173, 177, 178, 180
    United States industry and 96–102
fuel efficiency 21, 26, 147

GATT 174–5, 186–7
General Electric (GE) 13, 125–6, 127, 138, 141, 143–4, 150, 151, 171
    government aid and 158–9
    Rolls and 14, 35, 143, 148–50, 152–3, 199
    Snecma and 29, 127–8, 130–7, 152–3, 199
    UDF and 137, 147, 186
    United States aircraft industry and 29
    see also, CFM International and CMF 56 13, 125–6, 127, 138, 141, 143–4, 150, 151, 171
Germany, Federal Republic of
    A300 and 53–4

A310 and 55–7
A320 and 58–62, 74
Airbus Industrie and 64–5, 73,
    80, 84–5, 97, 98, 99, 164–5,
    195
aircraft industry and 58–61
collaboration and 42–3, 202–3
export credits and 169, 177,
    178, 180
United States, cooperation with
    99, 100–1
Giscard, Valery d'Estaing 98, 100,
    204
GOCO 30, 159
Group of Six 99
Group of Seven 99
Groupément d'Intérêts Economique
    (GIE) 54, 65, 78–9, 167
Grüner, Martin 33, 42, 43, 55, 61,
    109, 161
Guiness Peat 182

Hall, Sir Arnold 201
Hawker Siddeley Aviation (HSA)
    34, 53, 54, 55, 97, 195
Hermes (German export credit
    agency) 169
Hermes (space vehicle) 202
Heseltine, Michael 203, 206
high technologies 4–5, 6–7, 205
Hochmuth, M.S 194
Holland 43–4, 107–13
House of Representatives *see*
    Congress
hyper-sonic transport 188

Iacocca, Lee 206–7
IATA 21, 25
Italy 58, 134–5, 168
Indonesia 44, 198
International Aero Engines (IAE)
    127, 128, 150, 151, 184, 199
CFM56 and 146–7
composition of 144
organisation of 145–6
origins of 138–45
see also V2500
International Lease Finance

Corporation 183

Japan 5, 9, 35, 197–8
Airbus Industrie and 114, 117
aircraft industry and 44–7
Boeing and 114–18
engines and 10, 127
Fokker and 44, 115
IAE and 143–5
MDD and 115
Rolls and 143–5
threat of 119, 143, 194
Japanese Aero Engine Corporation
    (JAEC) 45, 142, 144, 145–6
Japanese Commercial Aircraft
    Corporation (JCAC) 45, 117
JDAC *See* Japanese Commercial
    Aircraft Corporation
jet engine, impact of 19
JT3 29
JT8 131, 159
JT9 29, 53, 129–30, 138–9
JT10 30, 125, 128, 136, 139–42,
    144
juste retour 9, 78
Justice Department 139, 146

Keen, J.M.S 151
Keith, Sir Kenneth 35, 139, 140,
    171
Keyworth Committee 205
Kohl, Helmut 61
Kracht, Felix 83
Krook, Dan 21

Lagorce, Michel 161
Laker Airlines 169
Laker, Sir Freddie 57
Lamont, Norman 60, 80
Lathière, Bernard 59, 82, 83, 86,
    184, 196
launch aid 35, 58, 60–1, 62, 162,
    165
Layton, Christopher 38
Learmont, David 186
learning curves 18
leasing, aircraft of 181–3
Lee, Robert 179

Lockheed 20, 30, 54, 94, 129, 158
Lockheed L1011 20, 28, 29, 55, 94,
    158, 163, 171
Lufthansa 53, 61, 85
Lygo, Sir Raymond 36, 61

M45 108
MacFadzean, Lord 125, 127
Malroux, Claude 136
market, civil aerospace 17–26, 58
MBB 13, 41, 52, 61, 62, 83, 85,
    110, 160, 198
McDonnell Douglas (MDD) 13, 27,
    55, 81, 94, 97, 112, 161, 184
    BAe and 105–11
    collaboration and 94, 118–20,
      197–8
    Dassault and 98–101
    derivative aircraft and 58
    Fokker and 111–12
    government aid and 159–60
    leasing and 167, 182
    United States aircraft industry
      and 28–9
McDonnell Douglas, Sanford 99, 105
McKinlay, Robert 166
MD11 28, 166
MD80 27, 28, 58, 112, 167, 182
MD91X 185
MDF100 111–12, 115–16
Mercure 38, 96, 97–101
MITI 45–7, 115, 118, 142
Mitsubishi 45, 46, 144
Mitterrand, François 61
Mitterrand, General Jacques 61, 97
Morgan Grenfell 62
MTU 35, 41, 54, 127, 129, 130,
    138, 139, 144, 149, 150
Muehling, Odilo 127

NASA 135
Nau, Henry 7, 8, 9
Newhouse, John 47, 94, 105, 159,
    167
NGTE 142, 161

Office of Science and Technology
    205

Olympus, 37 128–9

Paillert, Pierre 184
Panam 19, 72, 77, 147, 171–2, 181,
    183–4
Panavia 43
Pearce, Sir Austin 59, 60, 80, 82,
    85, 165
Pearce, Joan 175
Phillips, R 22
Pierce, J.B.L 173
Pierson, Jean 83, 187, 198
Pratt and Whitney (P&W) 13, 35,
    125–6, 150, 158–9, 160
    collaboration and 119, 127–8,
      150–1
    export credits and 171
    Rolls and 138–42, 152
    Snecma and 37, 129, 131–2
    UDF and 186
    United States aircraft industry 29
    see also International Aero-Engines
predatory financing 177, 179, 183
Prill, George 175
Procurement Executive 70
propfan *see* UDF
protectionism *see* tariffs
Protsch, Otto 108
PW2037 *see* JT10

QSHA 95

Raeburn, A.R.G 143
Ramsden, John 110
RB 199 139
RB 207 53, 129, 130, 131
RB211 34, 54, 103, 125, 126, 129,
    136, 137, 140–2, 144, 148,
    153
RB401 139
RB432 142
Reagan, Ronald 160, 176, 205, 206
Reed, Arthur 33, 208
Revaud, René 136
RJ500 114, 142–3
Robbins, Ralph 146
Rolls Royce (Rolls) 13, 14, 22, 29,
    120, 160, 201

Airbus and 53–4, 55, 56, 130
BAe and 102, 108
collaboration and 35, 126–8,
    199–200
export credits and 171
GE and 148–42, 153
Japan and 142–6
P&W and 138–7, 152
Snecma and 128–31, 153
UDF and 147, 186
United Kingdom aircraft
    industry and 34–5
see also International Aero-Engines
Rowe, Brian 148, 149
Ruggie, J.G 8

safe harbor 176
SBAC 36, 39, 160
Schaefler, Johann 83
Schmidt, Helmut 106
SDI 12, 205, 208
Senate *see* Congress
Shorts 34, 125
Snecma 13, 29, 99, 127, 128,
    143–4, 149, 151, 201
  Airbus Industrie and 54, 135
  collaboration and 126–7, 199
  French aircraft industry and 37
  GE and 128, 130–7, 138, 152–3
  Rolls and 37–8, 108, 128–31,
    138, 153
  UDF and 137, 153, 186
  see also CFM International
South African Airways 55
Spain 40, 164
SST 20, 200
Stainton, Ross 103
state aid, civil aerospace and
    185–8, 200–2, 204–6
  Airbus Industrie and 85–6,
    163–8, 195–6
  Europe and 24–5, 157, 160–1
  export credits and 168–81
  France and 38–9, 162
  Germany and 41–3, 161–2
  importance of 1–3, 157, 159–61
  Japan and 45–6
  United Kingdom and 35–6, 162

United States and 30, 157–60
State Department 132
Strauss, Franz-Josef 61, 62, 67, 82,
    196
Sud Aviation *see* Aerospatiale
Swartouw, Frans 111

tariffs 24, 31, 33, 159, 174–5,
    186–7, 200, 202, 204–6
Tebbit, Norman 60, 146
technology, civil aerospace and
    17–20, 185–6
technology transfer 38, 74, 113,
    114, 130, 132, 139, 140,
    142, 143, 144, 146, 149,
    194–5
Thatcher, Margaret 36, 58, 60, 61,
    63, 203
Tokyo Round *see* GATT
Toulouse 71, 73, 82, 184
Transall 83
Treasury 61, 102, 140
turbo fan engine 20
Turbo Union 139
TWA 182
two-way street 10, 207

UDF (propfan) 28, 29, 115, 117,
    137, 147, 152, 153, 185–6,
    198
United Airlines 135
United Kingdom 1, 95, 108
  A300 and 52–7
  A310 and 56–8
  A320 and 58–63
  Airbus Industrie and 70–7,
    80–3, 96, 164–5, 196
  aircraft industry and 33–6, 162
  collaboration and 39–40, 202–4
  engines and 128–30, 140, 143,
    186
  export credits and 177, 178–9,
    180
  France and 39–40, 99, 101,
    128–31
  technology policy and 202–4
  United States and 39–40, 99,
    101, 128–31, 203–5

United States 1
  Airbus Industrie and 77, 119,
      183–5
  aircraft industry and 27–31
  collaboration and 93–6, 118–20
  engines and 130–42, 143–50
  export credits and 168–81, 187
  France and 96–102
  government aid to civil
      aerospace 157–60, 186–202,
      204–5
  technological challenge to
      10–11, 14, 206–8
  technological dominance of 7–8,
      10, 11
  United Kingdom and 102–7
USAF 158, 159

V2500 29, 45, 114, 115, 126, 127,
      136, 138, 144–7, 148, 184,
      186
VFW 614 108–10
VFW-Fokker 42, 43, 107–11
Viscount 19
V loan 158
Volvo 149

Westland 203
Williams, Roger 3, 7, 64
Worsham, J 28

Yoshinari, Takashi 47
YS11 45

Zysman, J 6, 11